Economic Analysis for Property and Business

Marcus Warren MA
Principal Lecturer in Land Management at De Montfort University

OXFORD AUCKLAND BOSTON JOHANNESBURG MELBOURNE NEW DELHI

Butterworth-Heinemann
Linacre House, Jordan Hill, Oxford OX2 8DP
225 Wildwood Avenue, Woburn, MA 01801-2041
A division of Reed Educational and Professional Publishing Ltd

A member of the Reed Elsevier plc group

First published 2000

British Library Cataloguing in Publication Data
Warren, Marcus
 Economic analysis for property and business
 1. Real estate business – Economic aspects
 2. Construction industry – Economic aspects
 I. Title
 333.3'3

ISBN 0 7506 4632 2

Library of Congress Cataloguing in Publication Data
Warren, Marcus.
 Economic analysis for property and business/Marcus Warren.
 p.cm.
 Includes bibliographical references and index.
 ISBN 0 7506 4632 2
 1. Economics.   2. Land use.   3. Real property.   I. Title.

 HB171.5 . W243
 333.33'2 – dc21   00–036024

Composition by Genesis Typesetting, Rochester, Kent
Printed and bound in Great Britain by Biddles Ltd, *www.biddles.co.uk*

Contents

Acknowledgements

My sincere thanks go to Kerry Marriott, Jane McCormack, Dr John Ebohon and Sonia Deadman who all contributed greatly to making this book possible. I would also like to dedicate the text to the memory of Professor Neil Bowman.

Introduction

This textbook has been carefully formulated to provide an easy to understand and highly practical examination of applied economics related to both property and business. Specifically, its main focus is a comprehensive examination of a wide range of property markets against the backdrop of how businesses can utilize property to its full potential in terms of both its use and its investment value. The reliance upon the inclusion of applied examples should enable the reader to appreciate the great scope that a basic understanding of economics can give them in their future academic and professional lives.

The fundamental strengths of this book primarily lie in its simplicity and the way in which it has evolved. With respect to its simplicity, it enables the reader quickly to gain a workable understanding of economics and economic analysis so that they can utilize the principles contained within. Such application can be in the form of enhancing an understanding of activity in a particular market, or in improving the comprehension of a report or research in a specific topic area. Moreover, it should also enable the reader effectively to influence matters themselves by applying the techniques learnt. Importantly, the text has been designed to be of use in any part of the global economy and its scope is therefore not limited to one country alone. With respect to its design and content this work has essentially been developed from a previous text by the same author and publisher, namely: *Economics for the Built Environment*. Recognizing that the primary purpose of the process of evolution is essentially to improve and adapt, subject to changing criteria, it can be seen that in exactly this way, this text has been modified to reflect changing economic conditions and thought. In addition, it has been improved and refined by many years of user feedback on the original work.

Essentially, the book aims to achieve its goal of being a workable text by setting itself a range of achievable objectives. These fundamental objectives are to enable the reader to:

- overcome any 'fear' of economics
- appreciate the practical relevance of 'theory'
- provide a logical framework for analysis

- effectively use the principles of economics to understand a wide range of applied issues such as:
  - activity in property markets
  - the behaviour of firms in business
  - the implications of government policy
  - the importance of changes in the world economy
- utilize the principles of economics to their own advantage in their educational, working and home environments.

Hopefully, this brief work will also go some way towards encouraging the use of economics to help overcome the many complicated problems facing the property business. Once a fundamental understanding of the discipline has been grasped it can be appreciated that it can provide a highly diverse and flexible analytical framework with which to examine applied issues and solve problems. In other words, fully informed individuals looking at the business of property need a thorough grounding in economics just as much as one in subject areas, which may initially be viewed as being more directly applied, such as property valuation and building performance to name just two examples. Moreover, the reader should look for, and appreciate the linkages between, such related subject areas rather than treat them as separate entities. To illustrate these points, this text will show, amongst many other things, that the type of buildings built, the technology and processes that are used to construct them, the materials that are used, the clients who occupy them and indeed their location, are all, to a large extent governed by, or are reactions to, economic decisions or are the result of more general elements of economic activity.

Who the book is designed for

This book would benefit all those who require a practical understanding of economics. However, it has been specifically designed for the following groups:

- Students of degree and diploma courses that are tailored to the requirements of gaining entry into the professions related to property and construction. Examples of such courses include Building Surveying, Land (or Estate) Management (both Rural and Urban), Quantity Surveying and Construction, to name just a few.
- Students of business who require an appreciation of land and buildings as part of an investment portfolio, as well as an understanding of the need to manage property to enjoy its full use potential.
- Students examining property issues in a subsidiary subject on non-property related courses. For example, many Economics degrees, or degrees in Combined Studies, have options looking at matters such as housing, the urban transport problem, the location of business, and environmental issues, all of which feature in this text.

- Practising professionals could also benefit from this book. The text clearly demonstrates that economic theory can be readily and successfully applied to examine real-world issues. In this way, for example, it shows why markets behave in the way that they do and suggests how they may change in the future. Moreover, people in business could obtain an enhanced appreciation for the reasons for the structure of their own industry and how such a structure influences matters such as profitability and output.
- Finally, many courses that are not related to property may find this a useful text simply because of its high degree of application. The level of application should enable the reader quickly to see the relevance and flexibility of economic analysis for the study of any industry whether or not it is in the world of property.

How to use this book

A textbook is essentially a learning tool that is designed to assist an understanding of a particular subject area. As with all tools, the full potential of this book can only be realized if it is used properly. As such, it is strongly recommended that you take great care in reading and absorbing this introductory section. Therefore, please avoid the temptation of moving directly to the first chapter, as this is analogous to building the ground floor of a building without providing it with a firm foundation. Such a building will be lucky to remain standing, and even if it does it is likely to suffer from substantial defects! Using the same logic, the book is at its most effective if the chapters are read in the order that they appear, as although the chapters are all inter-related and can be used in isolation, there is a logical progression and reinforcement of principles as the work develops.

It is also important to realize that this text has been developed as a companion, albeit a very comprehensive one, to additional information that can be gathered on courses via formal lecture, studio, and tutorial contact. Furthermore, and most importantly, the structure of the book is designed to give you an initial understanding, and appreciation for the use, of applied economic analysis by providing a basic analytical framework. To enhance your knowledge and understanding of a particular topic in more depth it may be advisable for you to seek out more detailed theoretical information from one of the many mainstream economics texts as well as specific readings in that area. The text is also designed to encourage you to further the analysis by obtaining data from your own country or area of interest in order to test the accuracy of its descriptions and predictions. Therefore, there is additional research to be done in order to expand upon the deliberate confines of this work. By leaving it to the reader to search for up-to-date and relevant data it ensures that the principles in the text can be applied over time and can be used for any case study in any location. In addition, it encourages you to become aware of the great range of data sources and information that are now available in the fields of business, property and construction, especially on the World Wide Web. To continue with the construction-related analogy used above, the book essentially provides a foundation and frame for a building, but other parts need to be added to the building in order to make it complete.

To ensure that you have understood and learnt the principles that are introduced within the book it is recommended that the following guidelines are observed:

- Firstly, read chapters as a whole at your normal reading speed. This should enable you to get the general gist of the argument, and introduces you to any new terminology. Then, re-read them at a slower pace, in a methodological manner, so as to understand their fuller logic and content.
- Make brief notes as you re-read the text.
- Re-draw any diagrams and try to understand how they work and how they could be used or adapted to illustrate different circumstances.
- Do not worry if you have to spend long periods of time understanding any one particular point or section. Never go on without understanding preceding information. It is important to *understand* what you read rather than memorizing it.
- Make a glossary of technical terms. This can be useful as a point of reference, but the exercise will also help you to remember key terms.

In order to gain maximum benefit from this text it firstly has to be understood what economics is and how its analytical framework can be of use. For many, the initial task is to overcome any fear of the subject area.

The methodology of economics

This brief section of the book is designed to give an insight into the methodology behind the subject of economics. It is by no means a comprehensive debate on scientific thought. However, by appreciating the general processes used in economic analysis, common misunderstandings will hopefully be eliminated before one tackles the substance of the text.

Essentially, economists have attempted to adopt much of the rigour of the traditional sciences. Such an approach allows for the development of theoretical frameworks that encourage the pursuit of consistent and thorough applied enquiry and analysis. In other words, economics should simply allow one to look at and solve problems in a logical manner. For example, if we wished to study the behaviour of a particular property market the following research methodology would provide the basis for a coherent examination of the issue.

Firstly, observations need to be made from the market so that ideas can be formulated about what useful research can be undertaken. It may be found, at this early stage, that there are so many issues or influencing variables to examine that in order to simplify the work certain realistic assumptions can be introduced. If some simplification is not introduced the resultant model or theory could be overburdened by information and by definition be cumbersome and unworkable. Such assumptions can exclude information that is not of great importance, or can assume that other, more consistent, variables are constants. This simplifying procedure may make a model marginally less accurate but it may well be the only way to make it useable at the local level or between markets. Conversely, the assumptions introduced should not be so extensive as to make the theory overburdened with constraints that make it effectively meaningless in the real world. However, as long as the end-user is aware of the initial assumptions made, these could easily be changed, or relaxed, if faced with notably different circumstances. In other words, before adopting a model in its current format the user should carefully examine its initial assumptions to clarify whether it is applicable for its proposed use.

Once these considerations have been made and this stage has been completed, an economic model, or theory, can be designed. The resultant theoretical model should not only be able to explain the current and past workings of the property market in question, but should also be capable of predicting the future of that market if exposed to stimuli such as a change in interest rates for example. The ability of the model to achieve these objectives should continually be tested to see whether it needs modification or complete replacement. Tests could be carried out by continually inputting data and examining whether the model predicted the correct results.

Data then needs to be collected to feed into the model. Such data can be collected at first hand (primary data) or obtained from already published sources (secondary data). In both instances extreme care needs to be taken to ensure that the data used is both correct and sensible. If an inquiring mind is not used at this stage the model will certainly conform to the well-used adage of 'rubbish in, rubbish out'.

After appropriate data collection the results of the model should be displayed, with the aid of descriptive statistics, in a clear, manageable and understandable form. Having reams of poorly summarized and collated data will be of little use. Again the decision may have to be made concerning the potential advantages of summarization. In this case the question is which data is important and should be included, and which is not and should therefore be left out. Moreover, when examining results, care must be taken to ensure that although descriptive statistics may promote an understanding, they can be, and often are, misused so as to create a particular emphasis. After this stage has been accomplished, the results need to be assimilated and analysed so that the model can be used in a meaningful manner, and can, for example, be used to forecast the future of a market given a likely set of impending circumstances.

The important conclusion from this approach is that there should be a recognition that theory and reality are not two completely separate issues. Simply, as already stated, good theory should be able to explain the current and past workings of a market as well as being able to forecast changes in that market in the future. Those that perceive theory and reality to be divorced could be suffering from any of the following scenarios:

- They have failed to appreciate that theories are normally simplified so that they can be adapted for a wide range of uses, and are easy to use.
- They have not realized that many theories are in fact deliberately designed to show extremes, rather than actual cases, so that we can compare real-life issues against these extremes, especially if either extreme is being approached.
- They have been let down by theory in the past and have thus become distrustful of theory in general. Such a reaction is understandable as theories can break down and become less applicable if they are not altered with time as underlying circumstances change. However, it is often the case that theory has been wrongly used or simply misunderstood in the first instance.
- They may be operating in practice in ways not suggested by existing theory. This could mean that there is an alternative and better theory available that contains principles not yet observed by the existing theory. Although an alternative explanation is that some firms, although surviving, are simply not maximizing their potential due to an ignorance of known theory and practice.
- Firms may be operating along exactly the same lines as outlined in theory but are simply unaware that this is the case. Such a situation should be unsurprising as any theory will, in part, via its initial observations, be

governed by how these firms operated in the first instance. In other words, theory is partially produced by observing reality. For example, a theory about house prices may also take on board the behavioural attitudes of estate agents and house builders.

In addition, potential negative aspects of theory need to be recognized and responded to so as to limit or eradicate their impact as shown by the following points:

- Theory should be open to constant critical examination. This does not imply that faults will always be discovered and that current theory is wrong or needs replacing. However, if problems are found the process of theory and application can be improved over time. Specifically, if existing ideas and practice were never questioned society would be unlikely to evolve. Indeed new thought has not only gradually altered our understanding but has occasionally drastically changed our previous perception of how things work. As such, theories should lose their support if they no longer work very well as models to help explain and forecast. For example, an economic model designed to describe the housing market one hundred years ago is unlikely to take into account modern variables that would need to be taken into account if the model were to work today. Faced with changing circumstances, the decision needs to be made as to whether the life of an accepted and existing theory can be prolonged by merely adapting it to take into account any new facts, or if the whole concept should be fundamentally rethought. At such a juncture it should be remembered that continually adapting a theory could overload it, making it impractical to use.
- The success of any science rests upon the initial ability of scientists to separate their observations of what does happen from their views of what they would like to happen. For example, the temptation to misuse data in order to prove or disprove a hypothesis depending upon personal views and bias must be avoided. Indeed this is important even at the point of data collection where one must avoid selectively choosing data in order to help prove a point. Hypotheses should be tested to see whether they are right or wrong. Furthermore, one should not be held back from testing views merely because they do not conform to the current mould of thinking.
- As with other sciences, it would be ideal to gain observations via controlled experiments. Obviously such a feature is unlikely when dealing with something as complicated and unpredictable as society. For example, if interest rates were to change by a certain percentage, all other economic variables, in reality, can not be held constant so as to isolate the specific impact of the interest rate change upon consumer expenditure. Incomes may also have changed and even the time of the year can alter consumption via seasonal spending. However, in an attempt to approximate the conditions of a controlled experiment, economics temporarily assumes that all other variables have been held constant. The Latin term *ceteris paribus* is frequently used to show that this assumption has been made. Indeed, in the following sections of this book this point is assumed unless otherwise stated.

- Many criticize economic theory on the grounds that it will not work due to the individuality of human behaviour. Evidence does show that there are certainly strong or eccentric people who occasionally do not behave in the way one would expect. However, on average, human behaviour is relatively predictable. For examples, hot weather increases orders for outside patios and barbecues to be built, and more people make use of leisure facilities such as golf courses and swimming pools. Therefore, theory should be able to predict results within a 95 to 99 per cent level of confidence depending upon the specific matter under examination. In fact much of modern-day economics (neo-classical economics) assumes the notion of 'rational economic man' whereby it is assumed that most, on average, behave in the same way as others.

In conclusion, although economics is still perhaps in its relative infancy when compared with more established sciences such as physics, it still has a solid foundation of thought stretching back for hundreds of years. For today's property and business communities it is perhaps interesting to note that much of the current economic theory applied in practice, especially in relation to the workings of property markets, owes its origins to the works of Adam Smith (1723–1790). Likewise, fundamental land theories date back to David Ricardo (1723–1823). The works of John Maynard Keynes (1883–1946) are still highly influential in prescribing the ways in which economies are managed whereby successful economies are likely to lead to buoyant businesses and property markets. This list is by no means comprehensive but gives the reader a small taste of some of the great names behind the discipline of economics.

Part 1

Market analysis for the property analyst

1 Applied property market analysis: the theory of demand and supply

This first chapter is designed to give a thorough, yet easily understandable, working knowledge of market analysis. After a brief introduction, the fundamental theory of demand and supply is examined. This foundation is then built upon in an applied manner for a range of markets in the property arena in Chapters 2 to 7. A sound comprehension of market analysis should enable one to explain a whole host of property-related issues such as:

- variances in house prices or commercial rents over time and between location
- the design of buildings and their location globally as well as at the local level
- the volume of new building taking place
- the relationship between the price of development land and the completed building
- the impact of changes in technology on buildings
- the impact of government legislation on property markets.

The list above is by no means exhaustive and is merely a flavour of what is to follow, but in each instance it is hoped that the reader will appreciate how economic theory can be successfully applied to real-world issues.

The essentials of the market system

The concept of a market is far from new both in terms of economic theory and in practice. For centuries, people all around the world have traded goods and services in formal marketplaces. Indeed, many of the urban areas that can be seen today have developed around areas originally designated to sell agricultural produce from the surrounding rural areas. In such locations traders initially set up stalls to sell goods to the general public that visited that market. Today, some of these market areas still remain although much of the activity has dissipated into a variety of retail outlets. On the theoretical side, perhaps the best known advocate of the market system was Adam Smith who described the workings of the market as an 'invisible hand' in his famous text *An Enquiry into the Nature and Causes of the Wealth of Nations* written in 1776.

By looking back at the origins of markets it can be seen that farmers grew crops and reared livestock that were then sold on to market traders who would then, in turn, sell them on to the consumer perhaps in a modified form. The type of output that the farms produced would be largely governed by what people wished to buy. It was obviously illogical to supply products that nobody wanted as they would remain unsold, although occasionally errors inevitably occurred, i.e. too much or too little of a particular good could be supplied as fashions and tastes changed over time.

For example, imagine if a particular type of vegetable was very popular with consumers one year, farmers would be enticed to plant more of it in the next growing season. However, because of the length of time involved between planting, harvesting and distribution, by the time the new crop reached the market, the product's popularity may have dwindled. Such difficulties of time lags would be even more pronounced when looking at the commercial rearing of animals.

Moreover, supply could also be affected by variables beyond the farmer's control. Poor weather, or an unanticipated infestation of crops by pests or disease, could lead to a bad harvest. Conversely, better than expected weather conditions could produce a bumper crop. Thus, the farmers and the stall holders were a source of supply aimed at satisfying public demand although situations were likely to arise whereby the market was over-supplied or there were shortages of some items.

Those that visited the market were the source of demand for the goods on sale. The public attended the market to purchase directly for themselves and shopkeepers could purchase in bulk. Shopkeepers would then sell the produce on to the consumer through their shop or outlet, perhaps after modifying or improving the good in some way. The bargaining process that ensued between supplier and consumer dictated that if there was a lack of demand, the selling price of the good was likely to be driven down. This was especially the case if the good was readily perishable and needed to be disposed of quickly. Conversely, prices would be driven up if shortages occurred as people competed for the limited supply.

In this way we can see that markets are basically an interaction between the two forces of demand and supply. As a consequence the term 'market analysis' may be interchanged with the phrase 'demand and supply analysis' as they both refer to exactly the same process. The benefits of such a system are that it should operate in the consumer's interest by providing products that are desired by the public. Moreover, if there is sufficient competition the goods should be sold at reasonable prices. However, although the mechanics of the market system are automatic and should need little regulation, there are instances whereby the state may intervene on the grounds of health and safety and ensuring that there is fair competition amongst traders, for example.

In many countries, traditional markets such as that described above continue in existence. However, developments in international communications, trade and technology have enabled us to continue the market process of the interaction between supplier and consumer without the need for both parties to meet at a formal location. For example, the prices of products at a modern supermarket are determined by the availability of supply and the level of demand. However, systems exist that enable the consumer to order products through the Internet and have them delivered to their home without necessitating a trip to the physical retail outlet itself. Moreover, continuing with the agriculturally-based example of above, fluctuations in supply are more easily brought under control as we plant more disease-resistant crops, are able to preserve food for a longer period of time or make up shortages via imports from abroad.

As economies grew and became more developed through international trade and technology, the market system naturally encompassed a whole host of other goods and services. As such the subject of market economics was developed to enable analysts to focus upon how demand and supply theory could be used to investigate the operation of a diverse range of markets in the modern economy. By means of example and application this textbook examines a range of markets related to property such as those for housing, offices, retail outlets and development land as well as those for building labour and construction components. By reading on, it will be demonstrated that the fundamental principles of market analysis are the same in each instance and that a knowledge base in this area will provide an understanding of how markets have behaved in the past and how they are likely to behave in the future. Moreover, if you wish to investigate non-property-related markets your analysis should follow the same logical progression.

The strength and durability of the market system has been demonstrated by the success of those economies that have allowed it to proliferate, subject to sensible checks by the state in areas where it may fail. Indeed, the market system has been exported throughout the world as economies become increasingly interrelated through international trade. As such many countries that formally tried to control the price and output of goods and services through a state-controlled centrally planned system have now embraced the market system in many areas. However, it has to be said that no economy is a purely free market economy as there is always a degree of state intervention and provision. As such one may imagine a spectrum with the two extremes being the free market at one end and the completely centrally planned economy at the other. All countries will obviously fall within this spectrum, some closer to one end than the other.

Applied demand theory

As ordinary consumers we all demand a variety of goods and services on a daily basis. We require food, clothing and shelter to name the most fundamental of our basic needs. However, our demands will normally only be met if we have the purchasing power to back them up. Thus, economics is interested in measuring 'effective demand'. That is demand that results in an actual transaction taking place rather than simply examining people's desires or wants which are by definition less quantifiable, but may be of interest to other disciplines such as sociology. For example, many of us may dream of owning a large mansion house set in large secluded grounds. Such a dream is a desire or want that most cannot bring into fruition because of a lack of sufficient income to purchase such a property or to service a mortgage upon it. The reality is that we demand shelter, and the type of accommodation that we actually purchase or rent is our real, effective demand. Obviously real transactions can be measured so that people's behaviour can be analysed for a variety of purposes, as we shall see in the forthcoming chapters of this section.

In addition to consumers demanding goods and services, firms at home and abroad also have effective demands. Development companies, surveying firms or financial institutions have a demand for employees, office space, transport and materials, for example. In addition, one must not exclude an often significantly large government sector from the analysis, which, just in their provision of services at both the local and national level, is a substantial source of demand in the economy.

All of these sources of demand if summed together are referred to in economics as aggregate demand although it is normally more useful to break down the analysis into its smaller constituent parts. In conclusion, as economics is only concerned with effective demand rather than demands in general it is acceptable practice just to write or say the word 'demand' as effective demand is assumed.

Demand variables

Before undertaking any purchase of a good or service you are normally likely to assess a large number of variables that will help you in the decision-making process. Such assessment may be done automatically by your brain at a very high speed as you have become used to frequently making similar decisions. As a consequence you may not be consciously aware of computing and quantifying a range of issues before you come to a conclusion as to whether or not to buy. However, where such computation becomes more apparent is when either an unusual purchase is made or when one is dealing with the purchase of an expensive item such as a computer, car or house.

There are many variables that we are likely to consider. Economics attempts to both identify and quantify these so that interested parties can predict the purchasing behaviour of both individuals and firms. In this way, producers can hope to meet consumers' needs and thus make a profit. As a starting point in this exercise, a demand function can be drawn up which is essentially a measurable list of the most appropriate variables that largely explain the purchasing behaviour of the consumer, or group of consumers, in question.

In its simplest form the demand function can be written in notation form in the following way:

$$D = f(P, Y, S, T)$$

The letters designated for each variable are the normal abbreviations used in economics texts:

D = demand
f = is a function of
P = price
Y = income
S = substitute goods
T = tastes

Note that the letter 'Y' is used to denote income. This is due to the fact that the letter 'I' is used for investment.

A consideration of these variables would go a long way in helping to explain the purchasing behaviour of individuals and firms. Indeed, the variable 'tastes' encompasses a wide range of issues such as people's perception of quality, the importance of status, the potential for brand loyalty and even the impact of time. This simple function is easy to remember as the four chosen variables form a pronounceable word '*PYST*'. Remembering the word enables one to break it down again to reveal the initial for each variable.

However, more detailed analysis may require an examination of a wider range of variables in an attempt to explain more fully the reasons for demand and any potential changes in such demand. Thus, to extend the function, one could attempt to measure the importance of other influencing factors such as expectations, the impact of government policy, demographic issues, to name just a few examples. Which variables are the most important will depend upon the market in question. In order to isolate the impact of each variable they are examined in turn coupled with the assumption that all other variables are held temporarily constant. If this was not done the analyst may believe that a change in the market is due to the variable currently being investigated. However, this variable may be influenced by changes in other variables that are occurring at the same time. The term used for holding all other variables constant is *ceteris paribus*.

This chapter now goes on to discuss the influence of these variables in some detail so as to provide an insight into their relative importance. The discussion below will hold for the purchase or rent of any good or service. Examples of each type of transaction are used to demonstrate such flexibility.

Price

When considering the purchase of a service, such as a structural survey from a building surveyor for example, most consumers will initially investigate its price. Sensitivity to price is important, as the consumer first needs to know whether or not he or she can afford to undertake the transaction. In addition, few will want to pay more for a service than is required, and therefore cheaper options or providers are typically sought. In other words, quantity demanded is highly related to price. The more expensive

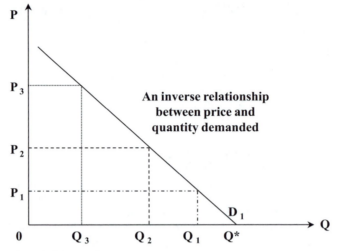

Figure 1.1 *A typical demand curve*

something becomes the less of it is demanded as people switch to substitutes or make do without the service. Thus, the 'law of demand' is often quoted as being: 'quantity demanded is inversely related to price'.

To help us envisage such a relationship, economics traditionally provides us with simple line diagrams as seen in Figure 1.1. This diagram depicts the general relationship between price and quantity in the form of a demand curve. In this case the line D_1 represents the given level of demand. Such a demand curve dictates that as price rises from P_1 to P_2 the quantity demanded falls from Q_1 to Q_2. Indeed, an additional increase in price to P_3 causes a further reduction in demand to Q_3. Thus, graphically there is an inverse relationship between price and quantity: as price goes up demand will fall, as price goes down demand will rise.

Using an example from the service sector, imagine if a firm of surveyors increased their fees for a structural survey from P_1 to P_2. The firm will find that, *ceteris paribus*, some people will now elect not to have such a survey conducted when they purchase a property, as it is too expensive. Thus, the number of clients will drop from Q_1 to Q_2. Therefore, there are those that will now take the risk of not having the survey done, although there will be some who have the work undertaken. The latter group ($0Q_2$) may require a structural survey in order to satisfy conditions laid down by financial institutions that are lending on the property. Such institutions would wish to have the security of the knowledge that their collateral or investment is physically sound.

Alternatively, the same relationship holds for a good such as housing. Simply, if a house builder increased the price of its houses, it could expect, *ceteris paribus*, a decrease in the number of people wishing to purchase its product. For example, increasing the asking price for a four-bedroom detached home in a particular area from P_1 to P_2 would lead to a decline in interest represented by the fall in quantity from Q_1 to Q_2, as people look to buy elsewhere or postpone their decision to buy.

Effectively, price changes will cause a movement along any given demand curve, whereas a change in any other variable will cause that curve to shift as can be seen by the illustrations below.

Income

In order to make a purchase in the market, money is required. Money can be derived from a store of wealth but is typically accumulated by receiving an income. Incomes can be received in the form of salary, rent received from property or dividends from shares. A consumer's purchasing power will be primarily determined by the disposable income that they have left after all deductions have been taken away. In its simplest form this will be gross income minus deductions such as income tax, national insurance contributions, payments to pension funds and so on. This can be expressed in notation form as:

$$Yd = Y - (Ty + Dy)$$

Where: Yd = disposable income
 Y = income from all sources
 Ty = income tax dependent upon the level of income
 Dy = other deductions taken at source that are dependent upon the level of income

However, it is likely that other deductions will reduce the amount of money that an individual has at the end of the day. For example, the consumer may need to make regular monthly payments to cover a mortgage and car loan, for example. It must also be remembered that an individual's ability to obtain a loan with which to make purchases is also dependent upon their credit-scored disposable income. As a consequence of being able to obtain borrowed finance the total potential purchasing power of consumers is likely to be greater than that suggested by their disposable income.

The ability of firms to demand goods and services rests upon the same principle. In this case their purchasing power is dependent upon how much profit they have left over after the total costs of production have been subtracted from total revenues received (see the theory of the firm in Chapter 8). Moreover, firms can also obtain loan finance depending upon the collateral that they can offer.

Imagine the scenario whereby the level of demand for retail units in a town was given by the demand curve D_1 in Figure 1.2. If the rent charged was typically as high as P_3 only Q_3 tenants would be interested in occupying such property. Other traders would either set up in business elsewhere or some may feel that it was not worth going into business in the first place if such high costs of occupation had to be covered. However, if there was an increase in consumer spending in the town this is likely to filter through to retailers receiving a greater income as they sell more output.

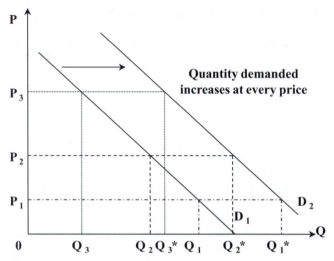

Figure 1.2   *Demand shifting to the right*

Higher incomes increase retailers' profitability and perhaps indicate that further benefits could be derived from expansion. Indeed those who had not originally located in the area may now be tempted to do so. The business plans of this latter group would now make more viable sense as high rents could be covered by high incomes. As such the demand for retail outlets is likely to increase. This is represented in Figure 1.2 by the demand curve shifting to the right from D_1 to D_2. Now it can be seen that more retailers are willing to pay the high rent of P_3 in order to acquire a shop. Indeed this diagram suggests an increase in interest by potential occupiers from Q_3 to Q_3*. Under the same circumstances of rising retail incomes and by using exactly the same logic, the demand for retail outlets would have increased from Q_1 to Q_1* if rentals had initially been as low as P_1.

Alternatively, an increase in the demand for valuation surveys will occur if an increase in incomes had fed through to increased activity in the property market. For example, if consumer incomes increased in real terms more people would be expected to purchase their own homes or move up the housing ladder. Therefore, if the ruling market price for a valuation survey was P_2 the number of people requiring this service would rise from Q_2 to Q_2*.

Substitutes

Most goods and services have substitutes that can be used instead of the one originally selected. Therefore the price of alternatives can affect the demand for any good or service. For example, imagine if the property managers of an office building increased rents from P_1 to P_2 as depicted in Figure 1.3. This diagram suggests that some tenants ($0Q_2$) may be happy to continue

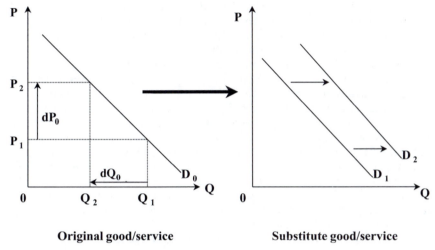

Original good/service   **Substitute good/service**

Figure 1.3   *The impact of a substitute good or service due to a change in the price of the original good or service*

occupying the building. However, others (Q_2,Q_1) would decide that the new rent was too high and move to an alternative building that met their requirements or cease trading altogether. As such the demand for a substitute good, in this case another office building, would rise. This is shown by a shift to the right of the demand curve for the substitute building showing that more people are now interested in occupying it than was the case before.

As well as substitute goods, one should also be aware of complementary goods. A complementary good is one that is consumed in conjunction with the original good. For example, by purchasing and owning a house a local property tax normally has to be paid as an ongoing yearly financial commitment. Domestic rates or a local council tax are both examples of such a tax. If these taxes were very high, or were substantially increased, in an area compared with other similar locations they may dissuade people from moving to that area. Therefore, the demand curve for properties would shift to the left representing a decline in interest for them.

Similarly, there is often a tax levied on the purchase of property such as stamp duty. Again this is complementary in nature, as a house cannot be purchased without paying for the tax. If this duty were increased from 1 per cent to say 5 per cent it may cause a reduction in the number of people who wished to become owner occupiers as the initial costs involved in the acquisition of property may be deemed to be too great.

Tastes

Consumer tastes are dictated by a variety of influencing factors ranging from brand loyalty to fashion and advertising. For example, a particular part of a town may currently be perceived as the best location for residential property. As a consequence, housing in that area attracts a high level of demand. Indeed, using the notion of a fashionable location as a feature in the advertising of a new development could attract further people to the area. As more wish to live there the demand curve shifts to the right. Using Figure 1.2 again it can be seen that despite high house prices of P_3, for example, the level of demand would increase to such an extent that now Q_3^* people were interested in occupying such homes rather than Q_3 as the demand curve shifts from D_1 to D_2. Conversely, at some future date, consumer interest could switch to another part of town altogether and the demand curve for properties in the original area would therefore move back to the left. Such a change in desirability could be due to the whims of fashion. Alternatively, there may have been a fundamental structural change in the local urban economy that encouraged a change in demand patterns. For example, the attractiveness of the original area may have been due to its proximity to a railway station that is then closed down. On the other hand it may have been adjacent to open countryside that has become built upon.

Expectations

The role of expectations of what may happen in the future is increasingly recognized as being a very powerful force behind demand. For example, if consumers believe that their incomes will rise in the near future they may increase their demand now rather than later. Such activity can be achieved by obtaining short-term loan finance that can be repaid when incomes actually rise. Thus, the demand for home improvements or new furniture may increase solely because of an expectation of rising incomes. Such an air of optimism will tend to shift the demand curve for most goods and services to the right.

Conversely, if consumers became pessimistic about the future they may curtail current demand so as to make savings now in case there is a negative change in their fortunes in the future. For example, people may fear an impending decline in their real incomes in times of high inflation, or even unemployment if economic activity were to slow down and a depression were to begin. Such beliefs are likely to cause the demand curve for most goods and services to fall. For example, plans for an extension to be built on one's house may be shelved for the time being.

Expectations of future increases in house prices may encourage people to buy a house now rather than later. Postponing the transaction could lead to the consumer having to pay substantially more for the same dwelling. Or, an inferior house may have to be purchased as other, more superior, houses go beyond the financial reach of the purchaser.

In a similar way, firms base their current investment expenditure upon forecasts of the future state of markets. This has to be done as new product lines and production techniques take time to install and the development of new premises takes longer still. Therefore, if industry is optimistic about the future they may increase their spending. Obviously spending would be reduced if they were pessimistic.

It should be seen that expectations could in themselves be self-fulfilling prophecies. For example, if there is a general air of optimism in the economy on behalf of both firms and consumers their spending will increase. As expenditure is increased demand levels rise. More is purchased from the shops, retailers will have to employ more staff and order more supplies from industry. As more goods are required, the manufacturing sector may need to hire more staff and so the process continues. In other words, a belief in economic growth has actually created it (see the debate on the multiplier process in Chapter 10). In terms of the housing example above, a belief that house prices will rise at some future date encourages present consumption. Thus, a rise in the demand for housing will cause house prices to rise.

It should also be recognized, though, that not all expectations are correct and come to fruition. Pessimistic views may have been unfounded just as beliefs in improving economic conditions could prove to be over-optimistic.

Government policy

A range of government policies at both the local and national level can influence the level of demand in property markets. We have already seen the potential impact of domestic property taxes in the debate above concerning complimentary goods. In a similar way the attractiveness of commercial property in an area can be dependent upon the level of business rates or whether there is any form of financial assistance on offer for new businesses. On a larger scale, government macro-economic policy is pivotal to decisions in the property market. For example, increasing interest rates as part of a deflationary monetary policy could reduce consumer demand, as loans become more expensive. Housing demand could also be adversely affected as the cost of mortgages rise. In addition to this there is likely to be a decline in investment demand from firms, as the funding of projects with borrowed monies becomes prohibitively expensive. For a full debate on macro-economic policy and property markets, please refer to Chapter 12.

In addition, government policy related to the environment can influence the level of demand for certain property types. The reinforcement of green belt policy or policy aimed at the protection of areas of outstanding natural beauty could increase the demand for residential properties on the periphery of such areas as these houses would be guaranteed the continued existence of a pleasant view. On the other hand, the demand for development land in the centre of urban areas may be increased if government were to offer financial assistance to firms willing to redevelop inner-city brown field sites. For example, redundant industrial property could be cleared to make way for a new apartment block.

Decisions regarding transport and its associated infrastructure obviously affect the demand for property. For example, the provision of a new motorway or rail link to a town could greatly increase that town's attractiveness to both industry and commuters. As such it would be expected that the demand for all forms of property would rise. More people would want to live in the area due to the ease of travel and as such the demand for residential property would rise. As more people moved into the town there would be an increased demand for retail outlets. Better infrastructure would also encourage a demand for office, industrial and warehouse use.

Demographic issues

Demography is essentially a study of issues related to the human population in general. It not only includes an examination of how many people there are and how this number may change over time, but also breaks down the total to examine specific sub-sectors. For example, demography looks into the movement of populations from one region to another, how many children are being born in each year and where they are born. It also examines the numbers of elderly and the longevity of the population. All of these factors can be vitally important in determining the level of demand for property in general, as well as for particular types of property.

For example, if there is a high birth rate in a particular region of a country planners would need to respond to ensure that a sufficient number of schools are built for all levels of education as those children grow up. In the same way house builders need to ensure that their developments contain properties that are attractive to families with children. Such housing may need to be designed with an appropriate number of bedrooms as well as offering facilities for a safe play area. Indeed, a developer may encourage demand by advertising properties that were designed to be child safe. As these children grow up there will be a demand for leisure facilities and for shops that cater for a younger age group. In the long run these children will themselves grow up and have children of their own. As a consequence the cycle of property demands begins once again.

Individual demand to total demand

It is often useful to investigate the demands of different sub-groups in a community so that the market caters for their diverse needs. For example, a house builder needs to be aware of the attributes of properties that attract different income groups. However, it is also useful to ascertain what the total level of demand is for most goods and services. For example, whether rich or poor, a valuation survey will still be required to be conducted in order to obtain a mortgage to buy a house. Therefore, a firm of surveyors may be interested in estimating the total level of housing demand at any given time.

Figure 1.4 could be used to illustrate the demand for housing from three separate communities. The elasticity of demand is different in each instance, which suggests that each group will react differently to a change in house prices (see Chapter 2 for a full explanation of the concept of elasticity). In this illustration, the demand curve D_A shows that demand would drop from 40 to 20 units if house prices increased from P_2 to P_1. Whereas the same price

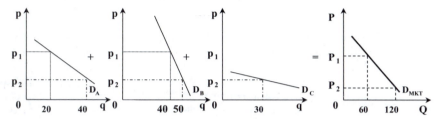

Figure 1.4 *Horizontal summation of demand*

increase would only produce a drop in demand from 50 to 40 units by those represented by demand curve D_B. In fact there appears to be no demand for houses at a price of P_2 in the case of demand curve D_C. These different responses can be totalled via a process known as horizontal summation to produce a market demand curve (D_{MKT}). Once this has been achieved the surveying firm would be able to estimate that with house prices as high as P_1 the total level of housing demand would produce 60 transactions. As such it is likely that this number of valuation surveys would be required. If interest in house purchase was stimulated by lower prices, say P_2, perhaps due to increased competition in the house-building market, the total demand curve forecasts that 120, rather than 60, surveys would be asked for.

Applied supply theory

The other side of the equation to demand is supply. Supply comes from a variety of sources in an economy ranging from an individual's willingness to supply his or her own labour to the supply of products on to the market by a large multinational corporation (see the debate on sources of supply in the economy in Chapter 10). All suppliers will essentially respond to a variety of stimuli that will encourage them to provide goods or services. Just as with the analysis for demand, a simple supply function can be formulated that highlights the main influencing variables on the supply side. This can be written in notation form as shown below. After this each variable will be discussed in turn.

$$S = f\,(P,A,S,T,I)$$

Where: S = supply
 f = is a function of
 P = price
 A = the aspirations or expectations of the consumer
 T = technology
 I = input prices facing the producer

Again, a simple word can be remembered to summarize the influences of supply, as supply is a function of '*PASTI*'.

Price

As with demand, price is perhaps the main driving force behind a producer's willingness to supply a good or service. If prices in the market are high it is likely that suppliers will receive higher levels of profit as each unit of their output is sold at a high price. Higher profitability will thus encourage greater output, as the rewards from production are clear to see. Conversely, if the market is depressed and prices are low it will hardly encourage suppliers to maximize their efforts. Indeed some suppliers may fail if prices are very low, as insufficient monies will be received for them to stay in business. This relationship between price and quantity supplied is shown in Figure 1.5. Here it can be seen that no supplier is willing to supply at a price lower than Px, whereas Q_0 output will be forthcoming if the market price were to rise to P_0. Higher prices will lead to greater levels of profit so that if prices were driven up to P_1 output would increase again to Q_1. This increase in output (Q_0Q_1) would probably be achieved by existing firms working harder as well as new firms joining the industry who were tempted by the promise of high returns. The actual level of the supply response would depend upon the elasticity of supply as explained in Chapter 2.

For example, if the commission paid to estate agents was in the order of P_0, Q_0 agents would offer their services. However, a boom in the housing market may increase these rates to P_1. If this were the case more work would be available for existing agents and new firms may set up in the local area. Thus the total amount of agency work would rise from Q_0 to Q_1. In the same way, a construction firm would be willing to tender for more contracts if the price received for each project was improved. Thus, in terms of the response of producers, the law of supply shows that there is a positive relationship between price and quantity supplied. As was the case with demand, any change in price causes a movement along the existing supply function. All

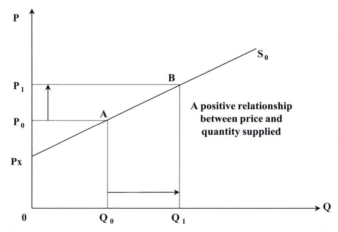

Figure 1.5   *A producer's supply curve*

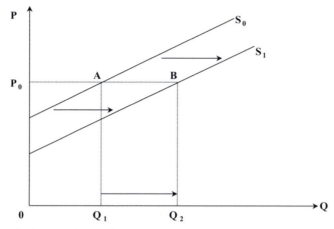

Figure 1.6 *An increase in supply*

other variables will cause the curve to shift either to the left or right depending upon the circumstances facing suppliers at the time as seen in the analysis below.

Aspirations

The aspirations, or expectations, of any producer is a key consideration in the supply function. The importance attached to this variable reflects the fact that most production decisions need to be taken in advance as they take time to come on line. As such producers need to forecast the future state of the market.

If producers are optimistic about the future and believe that they will be able to sell more at any given price they are likely to gear up production so that supply can be increased. This activity will be represented by a shift to the right of the supply function as depicted in Figure 1.6. Here, the supply curve has shifted from S_0 to S_1. This movement in the position of the curve implies that at a price of P_0 a quantity of Q_2 will be forthcoming rather than Q_1. Conversely, if they are pessimistic about the future and foresee declining sales they are likely to reduce output so that they are not left with unwanted and unsold stocks. Figure 1.7 represents such an eventuality as the supply function falls back to S_1 from S_0. At a price of P_0 only Q_0, rather than Q_1, will be forthcoming.

For example, if a house-building firm believes that interest rates are likely to increase and remain high for a period of time it may believe that the demand for housing will probably fall in the future. Such a fear of a decline in demand is likely to have its foundations in the view that higher interest rates may deter people from entering the owner-occupied market due to the fact that monthly mortgage repayments would be high. (For a further and fuller debate on this issue, see the analysis on the housing market in Chapter

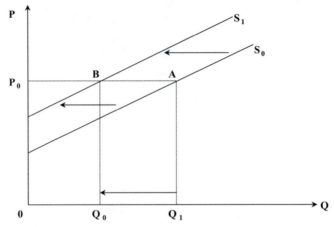

Figure 1.7   *A reduction in supply*

3 as well as that concerning the effectiveness of monetary policy in Chapter 12.) A belief in declining demand will reduce the number of new developments that are initiated by the firm.

Substitutes

Many producers are in the privileged position of being able to get involved in a variety of different forms of work. For example, a building firm, although specializing in a particular form of construction such as housing, could branch out and become involved in other areas such as building for the commercial property market. In the same way, given time to adjust, a firm of surveyors may be able to change its emphasis away from building surveying and towards planning and development if required.

As such, if one market recedes, producers can switch into a substitute good. For example, if the house builder found that there was a decline in the demand for houses in its traditional geographical development area it could switch to building elsewhere. Indeed it may move out of housing and into the construction of warehouses if the market for the latter was a more buoyant and therefore more profitable one. Moving out of the house-building market would cause a leftwards shift in supply as seen in Figure 1.7. In other words, at any given price this firm would build fewer houses as its main work is now elsewhere.

Technology

Technological improvements in the production process normally allow for more to be produced at a lower price through higher levels of efficiency. If this were the case suppliers will be able to supply more at any given price as they will be receiving the incentive of greater profits as the costs of production have been lowered. As with the diagram in Figure 1.6, Q_2 would now be produced at a price of P_0, rather than Q_1, as the supply function has shifted to the right from S_0 to S_1.

For example, a surveying firm that invested heavily in the provision of computers and laser-measuring equipment is likely to be able to undertake surveys quicker and more efficiently than a rival firm that has not adopted such technology. As firms modernize, the supply function would shift to the right as more surveys could be done without having to increase the price to the client.

It must be remembered that not all technological change is positive. Technology can be reversed if it is felt that such a reversal is in the interests of the general public. In the built environment new construction methods may be deemed to be unsafe on the grounds of health and safety if an increasing number of site accidents are reported. Therefore, their continued use may be prevented. Likewise, environmental concerns may prevent a new production process that is thought to be polluting the atmosphere or using potentially hazardous materials (see Chapter 7 and the debate concerning negative externalities).

Input prices

A huge array of inputs are required for the production of any good or service each of which has an associated cost (see the analysis of a firm's costs in Chapter 8). If input prices decrease, the producer is likely to be able to supply more at any given price and as a consequence the supply curve will shift to the right. On the other hand, a reduction in supply may come about in response to a rise in input prices. For example, wages may increase due to successful trade union action, or the cost of energy may go up in response to reductions in the supply of oil on to the world market. Again, imagine the case of the estate agency. If the costs of operating a branch office were to decline the agency may be in a position to open a new office to service another sector of the community. As such their output has risen in response to reductions in the cost of operation.

Individual supply to total supply

Knowledge of the supply behaviour of individual producers is useful as one can see how each firm reacts to changing market conditions. However, it is often the case that the total level of supply needs to be ascertained. In exactly the same way as various different demand curves were added together to produce the overall picture, supply functions can be horizontally summed in order to produce the overall supply curve in the industry in question. For example, the various responses of individual surveying firms as survey fees change over time could be assessed. These results can then be totalled to understand the reactions of the profession as a whole.

The formation of market equilibrium

One of the great advantages of the market mechanism is that both output levels and the price of that output is determined automatically by the forces of demand and supply without the need for any expensive forms of government intervention. Moreover, each time that there is a change in the market a new equilibrium is again found relatively quickly as the market adjusts.

For example, imagine the scenario whereby house prices on a new development were initially set too low as seen by a price of P_1 in Figure 1.8. With prices so low many would wish to purchase a property as represented by the high level of demand Q_{D1}. However, the returns to the developer are likely to be so insignificant that they are not willing to produce a large number of homes at this price. Therefore, supply will be as low as Q_{S1}. Thus, there will be excess demand as shown by the distance 'ab'. If people cannot obtain the house that they want they may offer the builder a higher price.

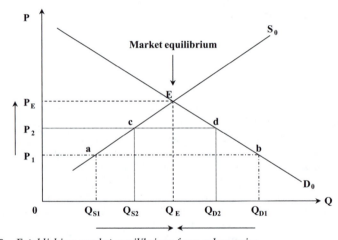

Figure 1.8   *Establishing market equilibrium from a low price*

Indeed, they may gazump another who has only offered P_1. As price is driven upwards the builder will have the incentive to build more houses. For example, QS_2 houses would be built at a price of P_2. As a degree of excess demand still exists at this price (depicted by the distance 'cd') prices again will rise until an equilibrium is reached at a price of P_E where Q_E houses are built and sold. At this stage there is no pressure for prices to change, as there is no excess demand.

If prices were initially too high as seen in Figure 1.9 the situation would operate in reverse. With a very high asking price of P_1 the builder would get few offers and much of the development would remain unsold. Specifically there would be excess supply equal to the distance 'hi' as demand would be as low as Q_{D1} yet the builder would wish to supply Q_{S1}. To sell the houses the

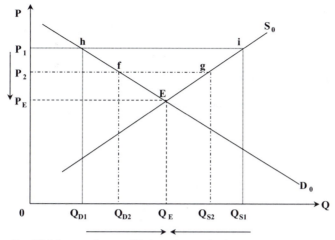

Figure 1.9 *Establishing market equilibrium from a high price*

builder would need to lower the selling price. As the price was lowered, more demand would be encouraged until we were once again at a market equilibrium.

The impact of changing market conditions

Although such market equilibria are automatically achieved they are not necessarily static. Both the supply curve and the demand curve tend to shift over time as shown in the discussion above concerning the variables of each side of the market. For example, if the level of demand were to decrease from D_0 to D_1 as seen in Figure 1.10 both price and output would be driven

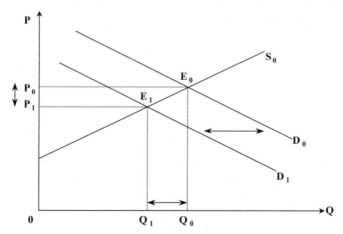

Figure 1.10 *The impact of a shift in demand*

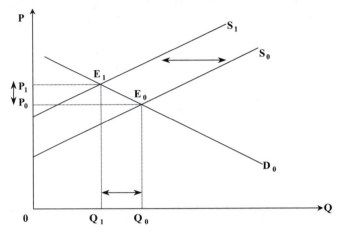

Figure 1.11   *The impact of a shift in supply*

down as the old equilibrium position of E_0 is replaced by the new one, E_1. (The opposite would occur if there was an increase in the level of demand.) For example, during a recession a surveying firm may suffer a reduction in the demand for valuation surveys as fewer properties are bought and sold. Fewer surveys will be undertaken, and to attract demand fees are likely to be lowered. Such a situation is depicted in Figure 1.10. Here, the number of surveys undertaken has dropped from Q_0 to Q_1 and the price of each survey has been reduced to P_1 from P_0.

Imagine if a major surveying firm left the market. This decision would reduce the supply of surveying services quite considerably forcing the supply curve to the left. Inevitably consumers would be left with less choice and thus the remaining firms would be in a position to increase their fees. Thus, as supply moves from S_0 to S_1, fees could increase their fees from P_0 to P_1 as seen in Figure 1.11. The opposite scenario would occur if a new firm joined the industry. This would shift the supply curve to the right and fee levels would be driven down, as firms had to respond to the increased competition in the market.

Conclusion

The market system is highly successful and widely used around the world. It has been developed over time to provide an efficient and automatic way of regulating economic behaviour. The next six chapters of this book concentrate upon a number of markets related to the property and construction industries. However, markets sometimes produce results that are not perfect. As such the state frequently steps in to modify the pure market outcome as seen in Chapter 6. Indeed, the market can, on occasion, completely fail as discussed in Chapter 7.

2 Measured market analysis – the concept of elasticity

In some circumstances it may be sufficient for the property analyst to assess the state of a particular market in quite general terms. For example, a client may be satisfied to know that house prices are increasing, because of rising demand, but he or she may not be too concerned to know the exact rate of increase other than that it is of the order of say around 5 to 10 per cent. Such behaviour may indeed be fully justifiable in that the rate may change over a very short period of time. However, as investing in markets frequently involves large capital sums for both consumers and firms alike, a more measured investigation into markets is typically required in an attempt to provide more accurate information. It is hoped that such increased accuracy will facilitate a reduction in risk and increase the prospects of achieving desired profit levels. In an effort to achieve this goal, economics provides the analyst with a 'measuring tool' known as elasticity. It must be stressed, however, that although by using this technique every effort is made to provide precise results, variables often change without warning and may also do so in an unpredictable way. As a consequence predicted results are unlikely always to be completely accurate.

Acknowledging this risk, the concept of elasticity is still useful and is fortunately simple to understand. A resultant ease of application can be seen by appreciating that although there are three key types of elasticity, only one formula (with minor adjustments) and two general principles need to be learnt.

The three categories of elasticity are:

- price elasticity
- income elasticity
- cross price elasticity.

This text now introduces each of these in turn and demonstrates their application. Please note that some economists may calculate elasticity using slightly different formulae so do not be concerned if you find an alternative method being used elsewhere, as it is the accuracy of the answer that is important. The general formula chosen below satisfies the criteria of being easy to remember and apply. Moreover, as stated above, it can be adapted for all forms of elasticity measurement with only minor adjustments.

$$Ep = \left(\frac{\Delta Q}{\frac{Q_1 + Q_2}{2}} \right) \div \left(\frac{\Delta P}{\frac{P_1 + P_2}{2}} \right)$$

Where: Ep = price elasticity
Δ   = change in
Q   = quantity
P   = price

Price elasticity

Price elasticity, commonly abbreviated to the initials 'Ep', is perhaps the most useful and common elasticity measure in property market analysis. This form of elasticity measures the responsiveness of the quantity demanded, or the quantity supplied in a market, due to a change in the actual price of the good or service under consideration. More specifically, the measure informs the analyst exactly how much quantity is likely to change for each 1 per cent movement in price.

Price elasticity can be calculated by inserting observed values into a simple equation based upon the general formula given above, although a number of variants of the price elasticity formula can be found. These formulae range from those giving point elasticity (the value of elasticity at any point on a demand or supply curve) to those that measure arc elasticity (the average value of elasticity between any two selected points on a curve).

Price elasticity of demand

Price elasticity of demand can be calculated using the following formula:

$$Epd = \left(\frac{\Delta Qd}{\frac{Qd_1 + Qd_2}{2}} \right) \div \left(\frac{\Delta P}{\frac{P_1 + P_2}{2}} \right)$$

Where: Epd = price elasticity of demand
ΔQd = change in quantity demanded
Qd_1 = the original quantity demanded
Qd_2 = the new level of demand after a change in price
P_1 = the original price
P_2 = the new price

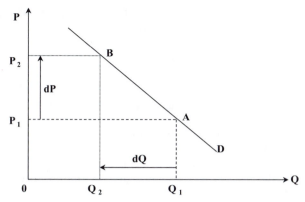

Figure 2.1   *Price elasticity of demand*

To help in the understanding of the workings of the formula and the meaning of its results it is advisable to refer to the following text in conjunction with Figure 2.1. The annotation on this diagram corresponds directly with the notation used in the text.

When analysing demand it will be found that all answers to this elasticity equation are negative values. Such results are simply a consequence of the fact that there is a negative, or inverse, relationship between price and quantity demanded. That is, when prices rise, quantity demanded has a tendency to fall. Likewise, as prices fall, quantity demanded normally increases (see the law of demand in Chapter 1). Such an inverse relationship can be seen diagrammatically in the downward, negatively sloped demand curve. However, as all the answers to the equation are negative, and due to the fact that we normally tend to think in terms of positive values, it is quite normal, or indeed convention, to disregard the negative sign altogether, or at least put it in brackets. For example, when calculating the price elasticity of demand for a good such as replacement windows for housing, the formula may produce an answer of:

$$Epd = (-)\ 0.4$$

Such an answer suggests that, between points A and B on the demand curve, for every 1 per cent change in price there will be a 0.4 per cent change in quantity demanded. Thus, if the supplier increased prices by 10 per cent he or she could expect a reduction in demand in the order of 4 per cent.

Price elasticity of supply

Price elasticity of supply can be calculated using what is essentially the same formula, as the same principles hold. The only alterations that have to be made are the insertion of supply, rather than demand, values. This marginally amends the formula to read:

$$Eps = \left(\dfrac{\Delta Qs}{\dfrac{Qs_1 + Qs_2}{2}} \right) \div \left(\dfrac{\Delta P}{\dfrac{P_1 + P_2}{2}} \right)$$

Where: Eps = price elasticity of supply
          Δ = change in
          Qs_1 = the original quantity supplied on to the market
          Qs_2 = the new level of supply after a change in price
          P_1 = the original price
          P_2 = the new price

With supply, one would expect all answers to this formula to be positive because there is likely to be a positive relationship between quantity supplied and price. That is, when price goes up, the incentive to supply more increases as more profits can be reaped. Hence a positively, or upwards, sloping supply function is observed when viewed in dia-grammatic form (see Figure 2.2).

 If a producer's ability to supply was being investigated, running appropriate information through the equation may give an answer that the price elasticity of supply is:

   Eps = 1.5

This answer indicates, for example, that between points C and D on this specific supply curve, the supplier is able to increase output by 1.5 per cent for every 1 per cent increase in price. Thus a 5 per cent hike in the market price would entice a 7.5 per cent increase in output.

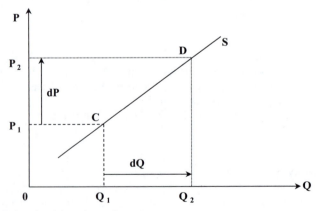

Figure 2.2   *Price elasticity of supply*

General price elasticity results

With respect to both demand and supply it should be realized that, apart from a few exceptions, the precise value of price elasticity will vary at different points along the respective curve. However, although it can be dangerous to generalize, the results should enable one to predict with a greater degree of accuracy how markets will react to changing circumstances. For example, knowledge of price elasticity should enable one to foresee how much house prices in any given area would change due to a known rising demand. This could be achieved by measuring the current elasticity of the supply of housing in that area and forecasting any likely deviations from the current figure (see Chapter 3).

Essentially the equations are likely to produce two broad ranges of result. Either:

- relatively elastic demand or supply; or
- relatively inelastic demand or supply.

Although there are certain theoretical exceptions to these two categories they are rare and are unlikely to occur in practice. However, they will be touched upon in this chapter as they give us extremes to act as useful benchmarks when examining the magnitude of either elasticity or inelasticity.

Relatively elastic demand or supply

The term 'relatively elastic' is used to describe any portion of a demand or supply curve where a change in price causes a more than proportionate change in quantity demanded or supplied. For example, relatively elastic demand suggests that a small increase in price will cause a more than proportionate decline in quantity demanded. In other words, the consumption of that good or service could be quite substantially reduced due to a relatively small increase in price. Exactly the same terminology and analysis can be applied to the supply curve. Relatively elastic supply implies, for example, that a small increase in price will encourage producers to supply much greater levels of output of their good or service on to the market. Therefore, under such circumstances, the response of the supplier is more than proportionate to the change in price. In either the case of demand or supply, a relatively elastic situation will be identified by the elasticity formula producing answers that are greater than one. In summary form this can be written as:

Epd > (–) 1

Eps > (+) 1

Such results imply that a 1 per cent change in price will lead to a more than 1 per cent change in quantity demanded or supplied. Assuming equally

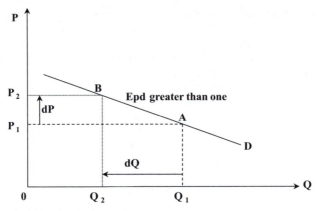

Figure 2.3   *Relatively elastic demand*

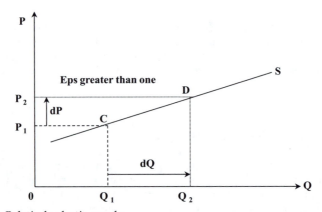

Figure 2.4   *Relatively elastic supply*

graded axes on a graph, the concept of a relatively elastic result is probably best envisaged diagrammatically as a relatively flat curve with a gentle gradient as illustrated in Figure 2.3 (demand) and Figure 2.4 (supply). However it has to be said that the actual value of elasticity will alter depending upon which portion of the curve is being examined. Both these graphs are designed to illustrate that small changes in price from P_1 to P_2 have led to correspondingly larger changes in quantity as depicted by the distance Q_1 to Q_2 in each instance.

For example, the price elasticity of demand equation could respond to data input into it with the following answer:

$$Epd = (-) 2.5$$

This answer shows that, for the specific price change that is being investigated, every 1 per cent increase in price results in a 2.5 per cent reduction in quantity demanded. One explanation of this outcome may be that consumers can easily switch to a substitute good or service if the original one becomes too expensive or non-competitive in price terms. (The latter part of this chapter examines the specific determinants of elasticity in some detail.)

Similarly, information gathered about supplier behaviour could be inserted into the formula to produce a figure for the price elasticity of supply such as:

$$\text{Eps} = 4$$

This result implies that, in the range of price change that is being examined, every 1 per cent increase in price encourages a 4 per cent increase in the amount of the good or service supplied onto the market. Such supplier reaction could be explained by the fact that the suppliers in question could easily increase output in the light of rising demand and prices, perhaps by being able to utilize and employ more unskilled staff, by making use of night shifts, or by enhancing productivity via better management.

The unlikely extreme of the elastic case would be a perfectly elastic curve where price elasticity is equal to infinity. This can be represented in graphical form by a completely horizontal line as seen in Figure 2.5. This figure illustrates a perfectly elastic demand curve. If a firm were to be faced by such a demand curve it would imply that as it attempted to raise the price of its product above P_1 it would end up selling nothing at all as there is no demand for this product at any price in excess of P_1. This outcome may come about because consumers could purchase a perfect substitute at a price of P_1, and therefore there would be no point in paying any price in excess of this price. On the other hand, if the firm charged less than P_1, say P_2, it would not be maximizing its potential revenue as more revenue would be obtained from selling an output of Q_1 at a price of P_1, rather than at the lower price of P_2.

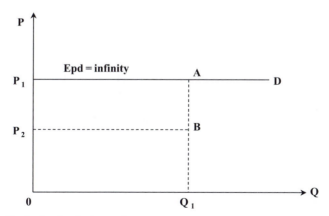

Figure 2.5 *Perfectly elastic demand*

Total revenue received by a firm is simply the amount sold multiplied by the selling price (P × Q). Therefore, it can be seen from Figure 2.5 that the revenue gained from selling a quantity of Q_1 at a price of P_1 is given by the area of the rectangle $0P_1AQ_1$. This area is obviously visually greater in size than the rectangle $0P_2BQ_1$ that would be received by selling Q_1 at a price of P_2. Although such an extreme is an unlikely occurrence in the real world, it is in fact a useful comparative analytical point, which enables examination of actual scenarios that approach this situation. Use of the concept of the perfectly elastic demand curve is used to examine the behaviour of firms operating in conditions of very high or perfect competition (see Chapter 9).

In conclusion to the debate on relatively elastic cases, the higher the result of the value of the equation the greater the degree of elasticity.

Relatively inelastic demand or supply

A relatively inelastic demand or supply curve essentially exhibits the reverse characteristics of their elastic forms discussed above. With a relatively inelastic curve a change in price will cause a less than proportionate change in quantity demanded or supplied. For example, an increase in price will only cause a small drop in quantity demanded in the case of demand, and only a small increase in quantity supplied in terms of supply. Therefore, all cases that exhibit inelasticity will produce answers to the price elasticity equation that are less than one. In summary this can be seen as:

$$Epd < (-) 1$$

$$Eps < (+) 1$$

Such results imply that for every 1 per cent change in price there will be a less than 1 per cent change in quantity demanded or supplied. Again this can be understood pictorially with the aid of explanatory diagrams by envisaging that, assuming uniform axes, an inelastic portion of a demand or supply curve is represented by a steep gradient, whereby the more inelastic the curve the steeper the gradient will be. Inelastic demand and supply curves are, for purposes of illustration, represented in Figures 2.6 and 2.7. These diagrams show a relationship between price and quantity whereby relatively large increases in price from P_1 to P_2 only induce relatively small changes in quantity from Q_1 to Q_2.

In such circumstances, data fed into the price elasticity formula could, for example, produce results such as:

$$Epd = (-) 0.25$$

This answer implies that, in this case, for each 1 per cent increase in price, quantity demanded would only drop by 0.25 per cent. This market response could be due to the good being viewed by consumers as a near necessity of

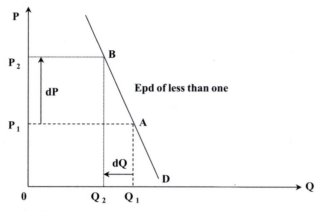

Figure 2.6 *Relatively inelastic demand*

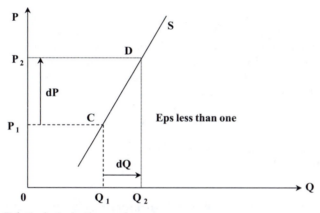

Figure 2.7 *Relatively inelastic supply*

life and as such consumption could not be reduced too radically in response to rising prices. The more inelastic the case, the lower this figure would be. In the same manner, when calculating the elasticity of supply, the figure would also be low such as:

Eps = (+) 0.2

This implies that for every 1 per cent increase in price, say due to rising levels of demand, suppliers only increase supply by 0.2 per cent, at least in the short run. Such an inelastic result could indicate an example of a good or service that is difficult or time-consuming to produce, and as a consequence changes in demand cannot be reacted to quickly. Just as with demand, the lower the figure given by the equation the more inelastic the case in question will be. The factors determining the degrees of such elasticity and inelasticity are examined at the end of this chapter.

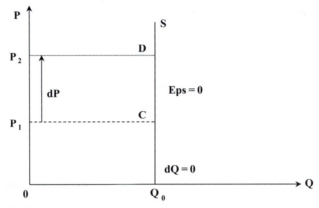

Figure 2.8   *Perfectly inelastic supply*

As with relative elasticity, the extreme case of inelasticity, in the form of a perfectly inelastic curve, is unlikely to occur in reality, but is analysed for purposes of comparative analysis. Figure 2.8 shows a perfectly inelastic supply curve whereby any change in price causes no supply response whatsoever. As quantity does not change, the equation will produce the result that price elasticity of supply equals zero:

Eps = 0

Chapter 4 examines the likelihood of such a result in relation to a debate on the supply of land.

Concluding comments on the measurement of price elasticity

The spectrum of possible results from the price elasticity equation, ranging from perfect elasticity to perfect inelasticity, is summarized in Figure 2.9. It is important to emphasize, though, that in reality one generally observes relatively elastic or inelastic situations, although it must be appreciated that the extremes can be useful points of comparison. Moreover, it must be realized that the actual figure for elasticity will vary at different points along any particular curve. For example, one point of a curve will exhibit unitary elasticity where elasticity is equal to one, but for the whole curve to exhibit unitary elasticity it would need to be a rectangular hyperbola. In other

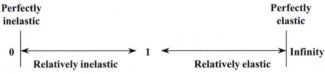

Figure 2.9   *The spectrum of elasticity*

words, every 1 per cent change in price would be matched by a 1 per cent change in quantity. Again, this represents an unlikely result, but is a useful comparative measure as unity represents the dividing line between elasticity and inelasticity.

Although a formula is available to strive for precise measurement and analysis, it is often not necessary to know the exact figure for elasticity or inelasticity, as simply a knowledge of whether a market exhibits either result will enable one to arrive at relatively accurate conclusions.

Factors influencing price elasticity of demand

The following are perhaps the key determinants of elasticity, although they should not be considered as an exhaustive list. Moreover, more than one of them can exert an influence at any one time.

The degree of substitutability and complementarity

Goods, services or assets that have substitutes tend to be relatively price elastic in demand as depicted by Figure 2.3. In fact, the higher the degree of substitutability between the items under consideration, the greater the degree of elasticity. In such circumstances the quantity demanded for the original item becomes relatively sensitive to a change in price as consumers can easily use an alternative if the price of the original item increases. If there existed a perfect substitute for the original item a perfectly elastic demand curve is likely to be observed, as seen in Figure 2.5. On the other hand, a lack of substitutes tends to give rise to a relatively inelastic demand curve as seen in Figure 2.6. In the case of an inelastic demand curve, even if the price rises for the original item, consumption cannot be varied greatly as few, or no, alternatives can be found.

With respect to goods, services or assets that are seen as complementary to one another, the degree of elasticity or inelasticity will depend upon whether or not the original good is inelastic or elastic in demand. For example, if the demand for the original good exhibits inelasticity, roughly the same amount of it will be consumed after a price rise than before the price change, therefore any items that are consumed with it will also only react with small quantity changes. Thus, inelasticity or elasticity of demand for the original good will correspondingly gives rise to inelasticity or elasticity of demand for the associated complementary good.

For example, bricks and mortar can be viewed as complementary goods to one another as one is of little use without the other. If the price of all types of brick were to increase, most builders would have initially to accommodate the price increases as they would still need to purchase bricks in order to build their buildings (see Chapter 4). In other words, their demand for bricks is relatively price inelastic. Such demand would only become more elastic if the builders either reduced their reliance upon bricks by changing the manner in which they build buildings by replacing brickwork with an alternative material, or they altered the design of the building itself. Note though that these options may not be acceptable from the point of view of the client for whom the building is being constructed. As a consequence, not wishing to affect potential revenues, it is likely that bricks will still be used in roughly the same quantities as before the price rise and thus roughly the same quantity of mortar will be required. In conclusion, inelasticity in the demand for bricks has led to inelasticity in the demand for the constituent parts of mortar.

The luxury/necessity syndrome

Logically, necessities are likely to be characterized by relatively inelastic demand as depicted in Figure 2.6. Indeed, the more essential an item the higher the degree of inelasticity, so that in the face of, say, price increases, the demand for such items will hardly alter. For example, the demand for energy to heat buildings in countries that suffer from cold winters is likely to be highly inelastic. If the price of energy increases, householders will still need to heat their homes to a comfortable and acceptable standard. Such continued consumption is normally made possible in the short run by the consumer reducing some other forms of non-essential purchase or by obtaining loan finance to cover the increased costs. A longer-term response could be to invest in energy-saving devices and components such as improved cavity or loft insulation or the addition of better glazing (see the section below concerning the 'time period').

Conversely, if a good or service is perceived as a luxury by consumers its demand is likely to be highly elastic as depicted in Figure 2.3. In such cases, if the good, or service, is not an essential for life, and it went up in price, consumers would be able to cut back on its consumption. For example, if the cost of non-essential extensions to houses, such as the building of conservatories, were to increase, the number of people demanding such work could easily drop without impairing the existing utility derived from the house in its present form. In the same manner the demand for expensive home furnishings could suffer the same fate.

The proportion of income spent upon a good or service

If only a small proportion of income is spent on any particular good or service, demand is unlikely to respond significantly to a change in price. As a consequence relatively inelastic demand is normally observed in such situations as seen in Figure 2.6. For example, if a house-building firm is faced with an increase in the cost of bathroom tiles from its supplier, the firm is unlikely to reduce its demand substantially as the price increase represents only a relatively small proportion of total building costs. In any case, bathrooms will still need to be tiled although putting fewer rows of tiles in each bathroom could make savings.

Broadness of category

When undertaking any research or analysis into property markets it is of paramount importance to define carefully the broadness of category that is being dealt with. Failure to achieve this invariably leads to the examination of highly misleading results. To demonstrate this point consider the following example: If the research area is defined very broadly such as 'residential accommodation' in general, highly inelastic demand will be observed, as irrespective of price most people will have to obtain shelter somehow. However, if one were just investigating a very specific type of housing in a certain geographical area, demand would be more elastic, as substitutes would be available.

The time period

The degree of elasticity of demand will tend to alter over time as people's reactions continue to adjust long after an initial price change. For example, when looking at urban transport problems, the policy-maker may wish to assess how the use of private cars is affected by increases in the price of petrol. If petrol prices were to rise the demand for private transport may initially alter only slightly as people find it hard to change their current arrangements. As such the use of private vehicles is likely to be characterized by inelastic demand in the short run. However, in the medium term travel habits could be altered by reducing the number of non-essential journeys or by organizing the sharing of lifts for example. Thus, in the medium term, the demand for petrol may well become more elastic despite an initial high degree of inelasticity. Indeed in the medium to long term some consumers may purchase more fuel-efficient vehicles as a response to this problem. Despite this demand behaviour, empirical evidence shows that even very large price increases are eventually forgotten in the long run as consumers become accustomed to the general higher level of prices. In such instances, long run consumption behaviour tends to return to the inelastic phase once again. This is an important conclusion for the town planner concerned with the provision of roads and car parking facilities in urban areas. Medium-term decreases in the use of private transport do not necessarily imply a long-term adjustment.

Factors influencing supply elasticity

A variety of factors will influence the elasticity of supply. However, the two most important determinants are perhaps the ease by which suppliers can change levels of production and their ability to react to changing market conditions over time.

The ease of response

The ease in which producers can adapt their equipment and production processes is a key factor in determining the degree of supply elasticity. If, for example, suppliers can react quickly to rising demand and prices their output will be elastic in supply (see Figure 2.4). Conversely, an inelastic supply curve suggests that any increase in demand will cause large increases in price yet little change in output. Such supply inelasticity will occur if the producer finds it difficult, or inappropriate, to increase output (see Figure 2.7). These conditions can arise when there are difficulties in obtaining a particular input or factor of production necessary to increase output. For example, a brick producer is likely to be able to increase the supply of bricks with relative ease in response to an increase in demand. This can be achieved, as the product in this case is relatively simple to produce subject to the availability of the necessary raw materials. However, at the other end of the spectrum, the house builder will not suddenly be able to increase output in response to a rise in the demand for housing. Suitable development sites need to be found, planning permission sought, and the product itself is quite complex. These are just a few of the characteristics of the good in question that are likely to lead to highly inelastic supply (see Chapter 3).

The time factor

When considering a very short time period the producer is unlikely to be able to respond quickly to any change in market conditions such as an unanticipated rise in demand. Changes in production may require organizing shift work for existing staff, taking on more labour, and ordering more materials. Indeed, at the extreme the firm may be operating at its fullest possible technological capacity. Thus, in the short run supply is likely to be inelastic. However, if output changes over a longer period of time were examined, supply will tend to become more elastic as the productive constraints mentioned above can be overcome. For example, in the early stages of a housing boom builders may initially be unable to supply the rising demand. Only in the long run will their supply become more elastic as new development sites come into fruition. Tempering such a response, however, house builders need to be wary of over-reacting to any rise in the market as demand increases could be short-lived and not an indication of a long-term improvement in the market. In the past building firms have obtained too much land, or have built too many houses that have subsequently been traded for lower than anticipated prices as the market slumps.

Income elasticity of demand

Whereas price elasticity of demand measures the responsiveness of quantity demanded due to a change in price, income elasticity of demand is simply a measure of the responsiveness of quantity demanded due to change in income. The logic is thus highly similar and calculations can be achieved with minor alterations to the general elasticity formula:

$$Ey = \left(\frac{\Delta Qd}{\frac{Q_1 + Q_2}{2}} \right) \div \left(\frac{\Delta Y}{\frac{Y_1 + Y_2}{2}} \right)$$

Where: Ey = income elasticity
         Δ   = change in
         Y_1 = original income
         Y_2 = new income

After the input of relevant data the answer to this formula is normally a positive figure. Such a result is due to the fact that most goods and services can be categorized as normal goods whereby there is a direct and positive relationship between quantity demanded and income. In other words, if your income increases you tend to buy more normal goods. If the formula arrives at a negative answer it implies that the good is an inferior good. In the case of inferior goods people tend to buy less of them as their income rises. An example of an inferior good that would exhibit such a response are economy home fittings such as flat packed self-assembly furniture. Cheap forms of construction, smaller buildings, or buildings in poor locations, may also be viewed as being inferior goods. For example, whether one considers the house buyer, or the commercial client seeking suitable business

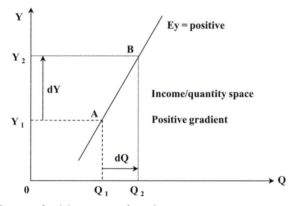

Figure 2.10   *Income elasticity: a normal good*

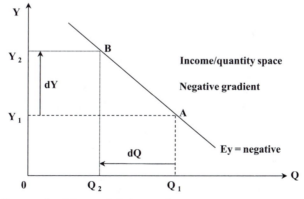

Figure 2.11   *Income elasticity: an inferior good*

premises, both parties may be forced to purchase, or rent inferior, low-grade dwellings in times of low income, or an economic recession. It would be expected that both would hope to move to more superior premises once the economic climate has improved.

The relationship between income and quantity demanded can be seen by referring to Figure 2.10 and Figure 2.11, whereby a positive gradient represents a normal good, and a negative gradient represents an inferior good.

Cross price elasticity of demand

Cross price elasticity of demand is a measure designed to examine the effect of a change in the quantity demanded of one good or service due to change in price of another related good or service. This measurement enables goods and services to be identified as being either substitutes or complements and the degree to which they fall into these two categories. Yet again the idea of cross price elasticity of demand is not dissimilar to the initial concept of price elasticity so that a variant of the original formula can again be used.

$$Ex = \left(\frac{\Delta Qa}{\frac{Qa_1 + Qa_2}{2}} \right) \div \left(\frac{\Delta Pb}{\frac{Pb_1 + P_2}{2}} \right)$$

Where: Ex = cross price elasticity of demand
Δ = change in
a   = good or service 'a'
b   = good or service 'b'

If two building materials or components were used in conjunction with one another they would be seen as complementary goods. This would be the case if a particular type of aluminium window frame required a specialized type of fixing to secure it to a building for example. Imagining a scenario whereby the fixings were to become short in supply and thus more expensive, a decline in their demand is likely to occur. As a consequence, there would be a corresponding decline in demand for aluminium window frames and an alternative such as plastic or wooden frames would be used as they became cheaper to install in total. The strength of such complementarity would depend upon whether or not these fixings were the only feasible securement method or if there were alternative techniques available. Therefore, in all cases where goods and services exhibit complementarity the formula will arrive at a negative answer as the relationship between the price of the complement and the demand for the other, original good or service, is an inverse one. The more the goods behave as complements the greater the negative value.

Conversely, the formula can indicate that two goods are perceived as being substitutes for one another such as plastic and wooden window frames for example. With substitutes there will be a positive relationship between movements in the price of one good and the quantity demanded of the other good. Thus, the equation will arrive at a positive answer under such conditions. For example, if the price of plastic window frames were to increase, the quantity demanded of wooden window frames is likely to increase, as more will be used, as they become relatively cheap. The more positive the answer to the formula the more the goods are seen as close substitutes.

The above definitions and analysis serves to show that general market analysis can be made more exact by striving to quantify the responsiveness of markets to changes in price, income, and the behaviour of other goods.

3 Dynamic property market analysis

This chapter demonstrates how market analysis can be applied to provide an understanding of a diverse range of property markets. Such a broad approach is used so as to demonstrate the versatility of the theory in explaining how all markets behave in practice when subjected to a variety of influences. The debates in this chapter are by no means fully exhaustive, although they are intended to be thorough starting points for further discussion especially when examined in conjunction with up-to-date data. No numerical data has been provided in the text itself as it is felt that such data would easily date and tie the analysis down to a specific country or geographical area.

Owner-occupied housing

The housing market is frequently a popular area of study and research. Such popularity is perhaps primarily due to the fact that as most live in houses of one form or other (whether we live in public sector accommodation, the private rented sector, or are owner occupiers) we can easily identify with the good in question. Moreover, an awareness of the housing market is especially keen in countries which have a very high level of owner occupation, coupled with relatively high incomes. It is in such countries that housing is not just considered as somewhere to live, but as an investment asset also. Furthermore, because of the general level of public interest in this market the media often selects housing as a key issue of debate, rather than the commercial and industrial property markets which are, perhaps wrongly, seen as being more removed from people's direct interests. The demand for the latter two types of property is a derived one, and as such the link with the general public is not as obvious.

The analysis that now follows is concerned with the owner-occupier housing market. This housing market can be examined in a relatively simplistic manner by using the elementary theory and observations made in the earlier chapters. With such knowledge the relative importance of the variables seen in the initial demand and supply functions should be able to be ascertained. Namely:

$$D = f(P,Y,S,T)$$
$$S = f(P,A,S,T,I)$$

However, for a full analysis of the owner-occupied housing market, as with most other construction markets, a much broader spectrum of issues and variables needs also to be investigated. To appreciate this point the supply and demand of owner-occupied housing is now introduced in some detail so as to see how the total market actually operates.

The supply of housing is primarily inelastic, at least in the short run. Thus, initially, even if there were large increases in demand, few new houses would be supplied on to the market. This lack of response is despite the fact that it would normally be in the interests of house builders to supply more housing, under such circumstances, so that they could reap the potential profits from selling more units in a buoyant market. Supply inelasticity is demonstrated in Figure 3.1 where it can be seen that supply inelasticity leads to large changes in price as a response to a demand change, yet small changes in the quantity traded. Specifically, an increase in demand from D_1 to D_2 has led to an increase in house prices from P_1 to P_2, yet the volume of houses traded on the market has only increased to Q_2 from Q_1.

The reasons for this supply inelasticity are twofold. If most building firms are reacting to the market, rather than basing decisions on forecasts of change, as in the case of the speculative builder, it will take time for those firms to realize that demand is rising in the first instance. Even after recognizing increases in demand, house builders may cautiously wait for some time before committing themselves to new development. Such a conservative approach is easily justified, as firms need to be convinced that movements in the market will be sustained and are not just short-lived temporary fluctuations. Even if the decision is made to build, building firms will need to acquire the necessary building land (in the absence of a 'land bank'), plan the development, seek planning approval, organize production, build the houses, market them and sell them. In fact, completion is likely to take place in phases, the first phase coming on line a good year or so after

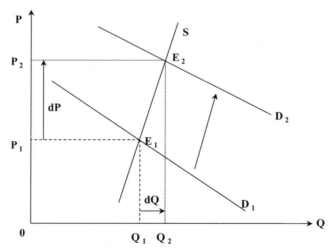

Figure 3.1 *The supply of housing in the short run subject to changes in demand*

project initiation. Thus, the first problem is that of a considerable time lag in the supply process of getting new output (houses) on to the market.

In addition it must also be realized that, in most countries, the existing housing stock is so large in the first instance, that any annual flows, or additions, to this stock are unlikely to be that significant in relation to the overall number of houses. Typically, additions to stock are usually represented by increases in the order of 1 or 2 per cent per annum. In other words the existing 'second hand' properties dominate the market and, moreover, the turnover in this market is frequently quite low.

Despite this inelasticity, however, new building does of course occur, and other additions to stock can come from a variety of other sources such as transfers from the rented sector, sales of public sector dwellings, as well as household dissolution and emigration. Builders can also improve their performance with respect to supply by attempting to forecast changes in demand in a more sophisticated way (the aspirations part of the supply function). Also, gains could be had from speeding up the actual building process itself. Suggested improvements in this context could be:

- Having 'off the shelf' plans.
- Quicker methods of construction (the technology variable in the supply equation). For example, more prefabrication could be used (it would have to be considered whether this could be done in aesthetically pleasing ways), and improvements in site labour productivity could be made (see the discussion on productivity in the construction industry in Chapter 4).
- Having stores of available building land (land banks).

The list could go on, although it must be recognized that such increased preparedness can be risky, and certain aspects of it will either cost money or represent a significant opportunity cost in terms of 'tied up' capital. For example, interest payments on borrowed monies to pay for a speculative land purchase could become very significant if the land is not quickly used, e.g. due to an unanticipated downturn in the market.

Linking these ideas in with demand, it can be readily appreciated that housing markets can be relatively volatile at both the local and national level due to demand fluctuations. As demonstrated in Figure 3.1, any changes in demand will cause large fluctuations in price. In order to understand why demand changes over time the basic demand formula can be investigated before going on to develop further, and more specific, variables applicable to this market.

Firstly, one would expect the normal price relationship to hold in the case of housing, giving a downward sloping market demand curve. Simply, at any one time, with given demand conditions, less people will effectively be able to demand houses when prices are high, and more people will wish to purchase when prices are low. High prices in the owner-occupied sector may force people out of the market into a cheaper rented sector for example. Moreover, at low prices some people may be encouraged to purchase more than one dwelling. Other houses purchased could be used as holiday homes or may be placed on the rental market as an investment.

Due to the high cost of purchase, another important variable to consider is obviously real incomes. Essentially, the higher real disposable incomes are the more money people have available for the acquisition of their own house. This is especially true where the ability to obtain a mortgage to help purchase property is income dependent. Normally, the total mortgage finance available for buying a house is based upon a certain multiple (say three times) of a borrower's annual gross earnings. For example, if incomes were to increase significantly in an area, say due to the prosperity of local industry or business, it is likely that more people could afford their own home, or indeed purchase other houses for investment purposes. Coupled with this is the fact that people on high incomes are likely to view rented accommodation as an 'inferior good', and thus switch into owner occupation as soon as they can afford to do so. To appreciate fully the impact of income as a variable it should be noted that despite short-run fluctuations, the house prices:earnings ratio in most countries is relatively static over time. For example, a house price:earnings ratio of 3.5 signifies that house prices in the country in question are normally three and a half times that of the average income. Generally, increases in incomes tend to lead to increases in demand for owner occupation of residential property.

The substitutes available are also an important consideration. If there is little in the way of public sector or private sector provision in the rented sector consumers may be forced into the situation of buying their own house. This point is especially true in countries where the extended family system does not operate and people do not tend to continue to live with their parents once they have reached a working age. When investigating the issue of substitutes one should not just identify the number of available rented houses or flats in the current stock. In addition there needs to be an investigation into the relative prices of such accommodation, their quality and their location. Each of these factors will be a key consideration when determining their real value as alternatives to owner occupation. The availability of real substitutes to buying one's own home takes away a lot of demand from the owner-occupied market as people rent accommodation rather than buy it.

Taste is an important variable also. In countries where home ownership is felt to be an important social goal, indeed often a reflection of social status, owner occupation is likely to be high irrespective of price. In other words owner-occupier demand exhibits a high degree of inelasticity. A desire for home ownership puts upward pressure on house prices as people compete with one another to acquire housing. Such taste-orientated demand can be enhanced further if there is a campaign, for example, induced by government, to create even greater levels of home ownership. Alternatively, house-building firms could attempt to attract demand via extensive advertising campaigns.

Obviously this initial set of variables is not an exhaustive one, and other factors need to be examined such as demographic change, and government policy at both the national and local level.

Under the heading of demography a variety of wide-ranging issues can be considered.

First, evidence and logic shows that housing demand is likely to increase dramatically about twenty to twenty-five years after a country has experienced a large increase in its birth rate. That is, demand increases occur after a country has experienced a 'baby boom'. This is because those born in the baby boom are expected to be in a situation to become 'first-time buyers' in their early twenties as they gain employment and start to have families of their own. Specifically this tends to cause heavy demand for small 'first-time buyer' dwellings thus putting upward pressure on the general level of house prices as existing occupiers take the opportunity to 'move up the ladder' to larger homes.

A second common demographic factor in many countries is an increase in the divorce rate. Marriage failure leads to more and more single people requiring single accommodation, whereas previously a couple would have lived in the same house. Linked to this is the modern trend to marry far later on in life. This trend can increase the demand for smaller housing units.

Another observable phenomenon in many nations is the rise in the number of elderly people, in particular elderly people living alone. This is primarily due to better living standards, improved medical technology, and the emergence of the nuclear family system. This results in many more houses being occupied by the elderly, and thus this part of the stock is not being released on to the market.

Finally, some areas, or regions, may experience changes in demand due to net immigration or emigration. For example, areas that are very prosperous, say due to newly-found economic prosperity, may experience increases in demand as more people move to the area to take advantage of its economic climate. Thus, huge increases in house prices will occur due to the likelihood of supply inelasticity. Conversely, of course, house prices may slump in areas where unemployment is high and people move away perhaps due to the closure of a declining industry.

Ruling, or anticipated, government policy is also a crucial area to consider. For example, general mortgage subsidies such as mortgage interest tax relief, or even renovation grants from the local authority may directly encourage demand. Just as importantly, although comparatively more indirect, is general government fiscal policy and monetary policy (see Chapter 12). For example, monetary policy determines the base rate of interest, which will have an impact upon the cost of a mortgage. If an individual feels that he or she can afford to purchase a house with a 95 per cent mortgage at an interest rate of 6.5 per cent, he or she may be prevented from doing so if interest rates progressively rose to, say, 12 per cent. Such an increase would greatly increase the level of monthly repayments as well as the total cost of the mortgage. Therefore, rises in the interest rate or, indeed, expectations of increases in the interest rate in the near future may well curtail demand.

In general, therefore, the level and interaction of demand and supply in that market determines the operation of the owner-occupied housing market. However, it is important to consider the effectiveness of the market system with respect to the overall allocation of housing. Such a debate is necessary, as it should never automatically be assumed that the market is the only or best solution. Indeed with housing, there is a substantial degree of public provision in many countries.

On the one side, commonly cited arguments in favour of the market system are:

- It should enable people to choose a house according to their own individual preferences, subject to the constraint of their income and their ability to raise mortgage finance.
- No expensive government machinery is required to build dwellings, allocate them, and manage them.

However, in reality the achievements of the housing market are potentially doubtful in as much that it is common to observe people unwillingly living in shared dwellings, or people living in accommodation that lacks basic amenities or is in serious need of repair. In fact, at the extreme, homelessness is still a problem even in the most advanced of economies. Thus, it would appear that a degree of market failure occurs. That is, the market fails to fully operate in the manner that is expected. Potential areas of such market failure in the housing market are touched upon below.

In times of rising demand the existence of supply inelasticity creates the potential of rapidly escalating house prices with little change in the existing stock of accommodation. This can reduce labour mobility within an economy, as people cannot afford to move from one area to another, more prosperous area, as house prices increase in advance of them. Moreover, it makes prices volatile so that owners cannot be too complacent about the future value of their home. In other words, if prices can rise rapidly they can fall rapidly also.

As housing is an expensive commodity few can purchase a dwelling outright and therefore most need to borrow finance. To cater for this, lending institutions, such as building societies and banks, typically offer loans for the purchase of housing in the form of mortgage finance. Inevitably, as all lending carries a degree of risk, these institutions tend to be understandably cautious and unwilling to forward loans to what they may perceive as high-risk cases. A lack of a prudent lending policy would obviously undermine the confidence of shareholders and depositors in the management of such institutions. However, some argue that excessive caution exists, and that this can lead to discrimination. For example, it has been known in the past for lending institutions to practise 'red lining'. One form of red lining was a policy of demarcating parts of a town or a city as areas of high risk, whereby there was a refusal to lend against houses in such areas. In fact, it has been claimed that such a policy of red lining was not merely geographical but was also used to discriminate against the ethnic origin of the loan applicant.

Another problem to consider is that of imperfect information. Some goods are purchased so often that consumers have a good idea of their ruling market price. Moreover, most retail outlets are in heavy competition with one another and therefore prices are maintained at roughly the same level throughout the market. If we did not have such (perfect) information some sellers could charge excessive prices and accumulate abnormal profits (see

Chapter 9 for information concerning perfect competition). However, imperfect information is widespread in the housing market, essentially because most people only occasionally purchase a house. Moreover, the costs of obtaining such information are considerable, for example time may need to be taken off from work for the purposes of 'house hunting'. New immigrants to an area are likely to suffer from imperfect information to a greater extent than local people as the former are, by definition, less aware of the local property market. It may be argued that estate agents can help to reduce this type of market failure by being the source of accurate market information. However, some argue that it is in the interests of the estate agents themselves to try to keep prices artificially high. This allegation is made on the grounds that as estate agents usually receive a percentage of the house price on completion of sale, in the form of commission, the higher the sale price the higher will be their income. If, for example, there are only a few estate agencies in a town they may attempt to interfere in the market to their joint benefit by secretly colluding together in order to 'decide upon' prices for the range of different types of property that are for sale. However, such behaviour, if it does occur, could 'backfire' on the perpetrators as housing demand declines in response to higher house prices.

The market may not fully take into account any associated externalities. An externality is created when the behaviour of one person, or group of people, can affect the welfare of another person, or group, other than through the price system. Externalities can be positive or negative. In the former instance benefits will be derived by some due to the action of others, whereas in the latter case the actions of some will affect others in a detrimental way (see Chapter 7 for a fuller debate about this issue). For example, if a householder lets his or her property deteriorate, due to a lack of repair and maintenance, it may become an eyesore and reduce the utility derived by other householders in the vicinity from their own housing. A neighbour continually hosting late-night, noisy parties could cause another negative externality. Such externalities may not be taken into account by the market mechanism. In the case of a noisy neighbour, for example, a potential purchaser may be unaware of this problem if viewing the property during the daytime. Although obviously not problematic, there are also potential positive externalities. For example, on purchasing a property one may be unaware that the people next door are planning to extensively landscape their front garden and improve the exterior of their house. Such activity could not only provide a more pleasant outlook, but it could actually enhance the value of other houses as the whole street is made to look more pleasant. This in itself could attract further purchasers and enhance property values.

It should also be recognized that housing is fixed in location and cannot easily be moved from one place to another. Although the movement of all structures is technologically possible it is unlikely to occur for reasons of common sense and finance and is thus an exercise rarely performed. Therefore, unlike many goods, shortages may persist in some areas whereas surpluses could occur in others. In other words, the market cannot rapidly adjust from one point of equilibrium to another after a change in the conditions of either supply or demand.

The market may fail adequately to cater for those on very low incomes who cannot afford to enter the market even on the 'bottom rung' of the housing ladder. Because of this problem it is often said that the market is capable of achieving an efficient solution but not an equitable one.

In conclusion it must be said that although there are problems with the market system it should not be automatically rejected as a suitable means of allocation. Rather one could:

- Link it with some form of government control or provision
- Find ways of reducing or even eliminating the sources of market failure discussed above.

For a detailed account of market failure in the built environment please refer to Chapter 7.

Technological change and the location of property

Since the beginning of human achievement, technology has largely shaped and led the way we live and how prosperously we do so. However, technological innovation has been far from constant giving us clearly identifiable 'leaps' in progress rather than a gradual, steady rate of growth. Numerous examples of such breakthroughs can be cited, ranging from the invention of the wheel to the development of modern space technology and the launching of satellites. It has also been a noticeable feature of human achievement that major pressures, such as periods of war, have given rise to the most rapid rates of technological growth. Improvements in aircraft technology and indeed the development of the jet engine owe much to the Second World War for example. It must be appreciated though that due to a variety of reasons, often economic or political, there are examples of technological reversal that have meant that some societies have suffered occasional set-backs rather than progression. For example, a country suffering from an energy crisis due to a lack of funds to import oil may not be able to produce sufficient and constant levels of electricity for business to make much use of computers that they had used when the economy was more buoyant.

Despite the possibility of potential reversals, it is fair to say that due to increasing economic pressures, international competitiveness, demographic growth, and the fact that one technological breakthrough may lead to another, the rate of technological progress over the past few decades has been staggering. In fact, advances are now being achieved at such a rate as to exhibit what could be described as exponential growth in many areas. Such changes are likely to radically alter the whole structure of how people live. In particular, 'revolution' after 'revolution' is occurring in the field of information technology (IT). Those businesses or communities that ignore or fail to keep pace with it may seriously jeopardize their competitive position and financial standing.

For all in the property world ranging from architects, construction firms, planners, and those involved in acquiring and investing in buildings, continual technological developments could inevitably radically alter the shape of the built environment and how people operate within it. It is relatively easy to identify a variety of broad technological changes that occurred in the twentieth century that substantially altered lifestyles and the environment. In industry, for example, there was the introduction of production lines and general automation that favoured long, low rise, industrial units in preference to the multi-storey factories of the original Industrial Revolution. In addition, the introduction of 'just in time' (JIT) technology went some way towards eliminating the need for the mass storage of industrial components and output. However, advances in information technology, although less direct in some cases, have had some equally, or even more, profound implications for all forms of property. Using the industrial example again, improvements in communications have ensured speedier contracts and have facilitated industrial liaison and assembly on an international scale.

Before expanding on these issues it is worth examining earlier developments in the field of information technology with the benefit of hindsight. This may assist one to forecast the impact of further changes that are occurring now or those that will take place in the future.

One piece of equipment that is common worldwide and has been in use for so long as to be taken for granted is the 'simple' telephone. Communication by telephone has obviously enhanced the speed of verbal communications both within firms and between them. Thus, it is an instrument whose importance should not be cast aside simply due to our familiarity with it or the perception that it is not an advanced piece of equipment. With technological advances via digital exchanges, international satellite communications, and other initiatives such as the mobile phone, the telecommunications network has produced a highly advanced and user-friendly aid to the business world at a competitive cost. With improved verbal communication generally and the ability for many to talk on the same line via telephone conferencing, for example, the telephone has undoubtedly produced progress and reduced the necessity for physical one-to-one contacts or even the use of a slower medium such as postal services. In relation to the latter method of communication the telephone has again given rise to advances whereby one has the ability to send written material through telephone lines via the use of facsimile (fax) transmissions. In terms of reducing the need for one-to-one contacts there are obvious savings to firms in that travel costs are saved as is time lost by having personnel travelling to appointments. On a national scale such a reduction in physical travel could go some significant way to reducing both traffic congestion and its associated pollution, both of which are important global environmental objectives (see Chapter 7).

Unfortunately, however, seeing the telephone as a pure advantage would be a simplistic viewpoint. Even today many parts of the globe are not sufficiently networked or catered for by telecommunication satellites to take full advantage of the technology. Indeed many communities, often in remote rural areas, have to suffer out-of-date and overused systems. Furthermore, some countries, especially those categorized as lesser-developed countries (LDCs) frequently do not have the wealth available to improve upon the situation. This lack of funding is perhaps partially responsible for creating a widening gap between the lesser-developed countries and the newly industrialized countries (NICs) as the latter are able to invest heavily in this field. Even in the industrialized world the telephone is often under-utilized as people are unaware of the huge array of facilities that a modern network can offer, or they may simply be uneasy about its use other than in its most basic form. Telephone conferencing, for example, is seldom used by many businesses despite its wide availability as are a number of other features such as preventing losing a client if one line is busy by having the call passed on to another line in the firm. Moreover, a lack of sufficient confidence on the telephone can be partially blamed for its under-utilization. Taking the negative arguments one step further, some see the telephone as potentially problematic as people are able to communicate directly without warning. This feature has been advanced yet further by the use of the 'fax' and the feeling that such communications require an urgent and immediate response. This point alone often necessitates a firm having to employ additional secretarial support. Furthermore, it is estimated that business loses time and money as staff misuse office lines for the making of personal calls.

Obviously computerization has been a leading force in the developments discussed above. Indeed, there is a great deal of interdependence here as many of the developments in the 'computer age' have been dependent upon the establishment of a highly sophisticated and reliable communications network. The use of computers though, and the ongoing growth in their complexity, capacity and user-friendliness has had far-reaching consequences at both the micro and macro level. Computers have enabled firms to shed large numbers of employees as certain tasks can now be done far quicker with the aid of new technology, or can replace the role of some staff altogether. Industries, such as the financial industry, who can and have readily adopted this technology, have seen an enormous decline in the number of personnel in recent years. At the extreme, branches can be closed with the advent of Internet banking. Computers, as long as they are adequately safeguarded, can also be relied upon to store information electronically that was previously kept in bulky, paper form. Not only is this development environmentally friendly but it can also save on enormous volumes of floor space, and thus rent, set aside for the storage of information.

However, just as with the potential under-usage of the facilities offered by the telephone the problem is often the same with computerization. In fact, due to the ever-increasing complexity of computers, this problem can be even more severe as suggested by a few common examples that now follow. First, at all levels within firms, there is often a lack of employee ability to use fully the capacity of the equipment made available. Rapid and ongoing technological advances in terms of hardware and software have resulted in many not managing to keep pace with such change. If this is ignored there are obvious problems of falling behind one's competitors, but continual retraining of staff is expensive in terms of both financial outlay and time.

Furthermore, some people who are at senior managerial level are, almost by definition, from a generation where such advanced equipment was not available and they are therefore unfamiliar with it to a large extent. This latter point can be identified as a common reason for the wrong systems being acquired for a firm, and often at greater cost, thereby preventing that firm from taking maximum advantage from the available technology.

There are firms that have yet to get sufficiently involved in this field perhaps due to failure to appreciate the long-term cost savings of computerization, cash flow shortages for the acquisition of equipment during times of recession, or fear that expensive new machines may themselves become rapidly out of date. A genuine fear and mistrust of new technology could further fuel a lack of investment in this area. Fear could arise from unfamiliarity, whereas mistrust could be founded upon the risk of technological failure or the ability of others to affect the firm's performance via fraud or the introduction of computer viruses, for example. Indeed, as with the telephone, some firms fear the abuse of their computer equipment as some staff may take the opportunity of access to browse the Internet for personal interest rather than for work-related reasons. Furthermore, a constraint faced within some organizations and certainly between organizations can be in the non-standardization of systems and software. This can delay communication, the retrieval of information and can often lead to staff needing to retrain as they move from one place of employment to the other.

On the positive side it must be recognized that great advances in space technology in recent times has meant that through the medium of satellite communication computers can 'talk to each other' on an international basis via the Internet, or World Wide Web, through electronic mail (e-mail). This inevitably enhances the potential of global communication and business opportunities.

Technological improvements tend to breed further advances as breakthroughs in one area frequently facilitate the ability of other areas to develop. Better computers with more capacity may allow for more versatile software for example. In this manner the growth of information technology has reached such a pace so as to exhibit exponential growth. As this process has the potential to drastically alter societies and economies alike it is perhaps fair to call it the 'New Information Technology Revolution' so as to distinguish this era from earlier slower developments, although ground breaking in their day. In terms of computers themselves, their growth in capacity, ability and 'user-friendliness' continues apace. Moreover, because of their incorporation into all levels of everyday life, be it at home or at work, increasing supply and growing competition amongst manufacturers has led to a reduction in the cost of equipment and better, more reliable products are now available. Examples of new computer technology are numerous and ever changing, but the following will help illustrate the importance of these developments.

Firstly, advances in microchip design and manufacture have enabled producers to supply highly powerful and versatile machines that are small and light enough to be carried in the form of a 'lap-top' or palm held computer. These machines, in conjunction with an ever increasingly advanced telecommunications network, have given birth to the concept of the 'portable office' and the ability of many to work effectively anywhere rather than in the physical location of a traditional office. Evidence of this can be seen today as an increasing number of people work in the comfort of their own home by engaging in 'teleworking'. To make life even easier and the technology available to all, voice-activated computers or ones that can recognize and convert the user's handwriting into formal text are replacing keyboards, and thus the necessity for at least basic typing skills. The implications for staff efficiency, employment and rental savings, as less employer space is required, is once again easy to see. Second, the introduction of 'CD' based machines has enabled the storage and subsequent access of greater volumes of material in smaller, less space intensive, formats. The consequences for information access and retrieval as well as storage space savings and their subsequent cost savings are again obvious.

There are still strong arguments in favour of the need for traditional one-to-one human contact in order to facilitate business. Such a view is supported by the fact that many deals are undeniably transacted or secured 'on the golf course' or over business lunches and dinners, or even at hospitality weekends. However, as a halfway measure, firms can communicate via the use of video links and video conferencing. Such facilities enable people to talk and see each other through a screen in an instantaneous manner. Such audiovisual equipment can be large, e.g. for the purposes of being installed in a meeting room, or the screens can be small enough to incorporate in the base of a hand-held telephone. Therefore personal communication is virtually achieved yet, at the same time, costs in time and travel as well as other related expenses are again reduced.

Many other technological innovations have taken place, are in the process of being developed further, or are being introduced in the near future. Such advances are rapidly turning the dreams of science fiction writers into reality. The growth of the concept of virtual reality is a graphic, and impressive, example. At first many dismissed virtual reality technology as suitable only for amusement. However, it has already proved to have had an enormous impact in industries where simulation rather than reality make good practice such as in the training of airline pilots or where extreme danger is possible, e.g. in the field of the production of nuclear power to name just two applications. The creation of cyberspace has enabled us to utilize a virtual environment in which one can communicate documents, arrange meetings, or even run exhibitions. As this technology becomes more available and acceptable to the user it can readily be introduced into a variety of far-ranging fields. For example, on the property front, an estate agent can now offer clients a preliminary viewing of properties within the comfort of an office via a virtual reality tour. Thus, rather than having to make many, often wasted, trips to sites the client can perhaps make a more informed short list in a far shorter period of time before finding the building that would meet their full purchase criteria.

The issues discussed above, as already briefly implied in some instances, will inevitably have far-reaching consequences for property markets. The impact of advances in information technology have already been felt in terms of rentals, capital values, the nature of the use of buildings, their disposal, as well as the planning and development of new buildings, and the way urban areas are designed and managed. In fact, many of the features of current property markets can be partially, if not wholly, linked to these events.

In terms of office property the implications are already implied and are all too easy to see. Modern computer and telecommunications technology has enabled firms not only to become more efficient and therefore reduce their labour requirements, but has also reduced their need for space. These two features alone result in less floor space being required which has a depressing effect on rentals via a lower demand for both existing and new buildings. Furthermore, if via the use of IT, further costs can be saved in terms of direct travel costs as well as time savings, a further loss of personnel is inevitable. Taking matters a stage further, especially with respect to increased familiarity and usage of video conferencing, the physical location of offices, other than in perceived prime areas, may become less relevant. In other words, if a company requires office space it may as well do so by locating in an area where office rentals are lower and skilled labour can be employed at a lower wage cost. Such moves can be further fuelled by exchange rate differentials that make labour of all categories significantly cheaper in some countries. Extending the issue further, allowing personnel to work from home, or out on site, rather than in a main office reduces space demand and thus rental costs. In addition the related running costs of providing office space are further reduced by effectively passing some of these costs on to employees at their home. Firms have achieved this by encouraging the notion of teleworking and 'hot desking'. In other words people are primarily based out of the office and only make occasional trips back to base to make use of a 'hot desk' that may be shared by others.

These developments are already well progressed, aided by a general public and government desire to avoid ever increasing traffic congestion and transport difficulties such as pollution and the risk of traffic gridlock. As a consequence, much of the floor space currently devoted to offices is unlikely to continue in its present format so that massive changes in use will be required to avoid a large number of derelict buildings in urban areas. This fear is specifically strong in those nations that allowed the rapid construction of speculative office developments during earlier boom periods of their economic and property cycles. It is already quite commonplace to find the location of the headquarters of multinational companies (MNCs) in lesser-developed countries. Or, companies locating in prime areas in the developed world have their administrative support abroad in order to take advantage of lower rentals, running and labour costs, as well as benefiting from any potential tax advantages.

As these changes occur, and perhaps at an increasing rate, they may provide partial solutions to some key problems of the day such as pollution and congestion. However, in turn, they throw open new difficulties and challenges to contend with. For example, property professionals need to decide what can effectively be done with unwanted office space so as to avoid dereliction and its associated social and opportunity costs, or indeed the wasteful demolition of such buildings. There is also the problem that if less people travel to a place of work, support industries and retail outlets in the vicinity of such land use will also decline leading to the blight of areas formally dominated by office use. To counteract this viewpoint to an extent is the belief that advances in IT have made firms more efficient and have thus encouraged consumption of a greater variety of goods and services. As a consequence technological change may have fuelled economic growth. Thus, according to this standpoint more firms have set up in business or have expanded in order to take advantage of increased economic prosperity. Therefore, a decline in office demand due to the introduction of modern technology itself may be partially offset by economic improvements created by such technology.

The spin-offs of these developments have inevitably filtered down to the residential property sector. As people begin to work from home, houses need to be designed with suitable working or office space as well as having the capacity to be wired up into the global telecommunications network. It is predicted that working from home may have long-term social con-sequences. On the negative side there is the fear of alienation and a lack of peer group motivation and contact. Moreover, productivity may be hindered by distractions at home such as being tempted to watch television or 'surf the Net'. Whereas, on the positive side, time saved by not commuting to work could well increase productivity as more time is released to concentrate upon a task. Indeed, the worker is likely to be in a better state for work having avoided the stress of travelling. Furthermore, reductions in travel time and travel costs may lead to the need for other forms of development in the area of leisure property as workers can afford both the time and money to visit complexes such as a gym or swimming pool for example.

In terms of industrial property, changes are also likely to be great, yet not similar, as normally the completed product still needs to be assembled on location. However, components can be made from home or at least in a variety of locations. An example of the former is in the production of some electrical items, whereas the latter is frequently exhibited in the aerospace industry. Generally though, advances in information technology will help encourage global communications and global products as national bound-aries are crossed by technology and made increasingly irrelevant in terms of trade. Another feature may be the more rapid development and improve-ment of manufactured goods as firms are able to utilize virtual reality to speed up product development and testing. As with office property, more and better output should encourage increased demand for industrial property from those organizations that can take advantage of new technology. Such demand would help to go some way towards offsetting the reductions in demand caused by the decline in many of the traditional heavy industries in the developed world. However, the physical location of the two types of industry is likely to be different, leaving derelict property in some areas yet the need to develop new buildings in others.

As implied in the analysis above, for retail property the indirect implications of changing technology are again far reaching. In the simplest of terms, if traditional office locations often found in the centre of urban areas (Central Business District or CBD) decline, retailing in those areas will also inevitably suffer a reduction in demand. This prediction is based upon the belief that as the number of office workers visiting a formal place of work is reduced, they are by definition not there to eat, drink and shop at lunchtime or after work. Thus, both retailing and leisure outlets are likely to experience a decline in trade forcing rental declines and closures. Such a process gives a further push to developing out-of-town retailing centres, near suburban residential areas, yet leaves government and local authorities the problem of trying to breathe life back into ailing inner city areas. Technology, though, is also having a direct impact upon the relationship between the retailer and the shopper. As consumers become increasingly used to working from home and are more familiar with using computers the advent of shopping through the Internet is now available. People can already shop for goods ranging from food and cars to aeroplane tickets. Thus, large distribution warehouses are being built which may eventually put a break or reversal on the development of super- or hypermarkets. Moreover, the need for other points of sale such as travel agents and car showrooms, to continue with the examples used above, becomes questionable as shopping habits inevitably change.

In terms of property held in the public sector, their buildings and plans for future space requirements will be faced by the same IT-led forces. The size and staffing of taxation offices and passport offices, for example, could be reduced via a greater use of computerization. In many countries the need to adopt modern technology and reap its benefits is further fuelled by pressures for public sector spending cuts. There is a need to become more efficient and cost accountable, as well as streamlining various parts of the sector, such as the post office, so that they may appear viable for eventual privatization.

In conclusion, it has to be said that a new and continuing IT-based technical revolution is underway in the world. It can be perceived as a threat as it will change our social structure in the same fundamental way as the dramatic changes brought about by the first Industrial Revolution. Moreover, if such technological developments do not increase economic activity sufficiently it will inevitably lead to labour redundancies and create unemployment. However, successful nations and firms alike are essentially those that manage to turn threats around so as to convert them into opportunities. Those that manage to do this will give themselves an ever-increasing advantageous gap against those who fail to change successfully. Imagine an intercontinental airline flying the latest jet aircraft competing with a company that insisted on flying a mid-twentieth century propeller-driven aircraft. The implications for the prosperity and ultimate survival of each firm are obvious. However, many firms seem to be missing the opportunities available to establish a technological foothold or lead. A simple lack of knowledge about what is currently available is partially to blame as many are under the false impression that these ideas are for some distant future date. Yet there are many other factors that are, perhaps unnecessarily, delaying the introduction of available information technology. Senior decision-makers may be from a generation that is out of touch with modern developments and are unaware of their potential; indeed many may feel threatened by their introduction or have a fear of change. Such reluctance could be based upon the grounds of job security, but also a loss of management control. On the financial side, managers may be fully aware of the long-term cost savings and advances in efficiency, yet they do not have the initial capital with which to make the investment. In relation to this point is the worry that a business could undergo significant expenditure only to find that because of rapid technological progress, the equipment purchased had been quickly and greatly superseded. On the practical side there are those that are resistant to change as they do not trust the reliability of the new technology or are afraid that such technology is more susceptible to fraud and related crime.

For those that accept the inevitability of the introduction of new technology and working practice, a word of caution to heed is that they must be careful to ensure that equipment purchased is appropriate to their needs. All too common are instances of some staff suffering from a lack of access or are lumbered with under-powered machines, whereas others have equipment with capabilities well beyond their requirements. On the positive side, firms could rapidly reduce their space and employment demands with obviously advantageous cost implications. Moreover, if a firm becomes more efficient and effective, or merely impresses others with its access to new technology, this could greatly enhance its revenue raising abilities. Remember also that the adoption of IT in an economy could help conquer serious environmental problems such as traffic congestion and its associated pollution, but at the same time may give rise to equally serious yet unforeseen social problems.

4 Market analysis for the factors of production in the assembly of property

This chapter examines the main factors of production that are required to assemble a completed development, including an investigation of the market for development land, the labour market for construction workers, markets for construction components and a debate concerning the role of the property entrepreneur as both property developer and manager.

The market for land (with specific reference to building land)

In examining the factors of production it has been traditional to treat land as a special case. It is often stated that as there is only a given amount of land in any country its supply is completely fixed, and can thus be depicted as a perfectly inelastic supply curve as seen in Figure 4.3. Therefore, according to this view, any increase in demand will not lead to more land being supplied, as no more land is available. With such an interpretation the only effect would be radical changes in land prices. This is contrasted with the situation of most goods and services whereby increases in demand will encourage producers to supply more of the good or service in question. However, although it is correct to say that the total amount of land in a country is fixed (apart from small changes due to erosion or land reclamation, for example), this is merely the stock of land and does not reflect the amount traded or the incentive to trade. The latter movements are the flow of land traded in the market at any given period of time.

Therefore, when looking at development land in particular, one should not expect there to be perfectly inelastic supply. Rather, there will be the normal upward sloping supply curve as shown in Figure 4.1. This supply curve implies that there may be people who own suitable land with planning permission who are willing to sell to construction firms when the price is anything in excess of P_0. Indeed if the level of demand causes land prices to reach P_1, quantity Q_1 will be traded. However, there may be many landowners who will only sell their land if offered an even higher price. For example, some may well wait until the demand for land has increased to D_2

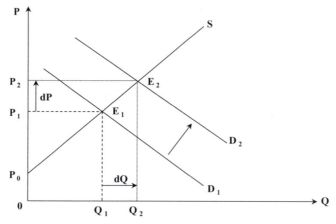

Figure 4.1 *The supply of building land and changes in demand*

giving rise to a new, higher, equilibrium price of P_2. This higher price of P_2 would encourage Q_1Q_2 more land to be sold on to the market.

Imagine, for example, that an individual lived on a one-hectare plot of land, and that he or she is allowed to subdivide the existing property into three further building plots as seen in Figure 4.2. Whether the individual does this or not is likely to greatly depend upon the price that he or she is offered for these building plots (assuming that the individual is not going to develop the land him or herself). Referring back to Figure 4.1, if the existing owner was offered a low price such as P_0 he or she would probably retain the property in its entirety valuing the space and seclusion that the land in its existing format provided. However, if land prices were to be bid upwards, say to P_1, due perhaps to general economic prosperity, or an increase in demand for housing in that area, the property owner may be

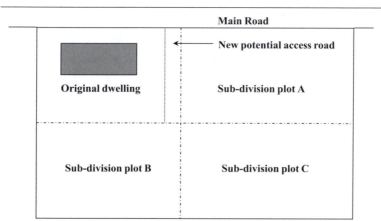

Figure 4.2 *The individual supply of land due to changes in price – a 'case study'*

tempted to sell off one of the pieces of land. Deciding which plot is to be sold may be determined upon grounds of the least loss of view, providing that adequate access and services could be provided to the sold land. Such a plot is likely to be either plot A or plot C depending upon specific preferences, as they are less in the line of view of the existing house. However, if the demand for building land increased yet further, giving a new equilibrium price of P_2, say due to a housing boom, the original property owner may be encouraged to sell off all of his or her available land. In fact it may be worth selling all the land and redeveloping the original plot in a more extensive and comprehensive manner rather than as a piecemeal development.

Thus, as with the productive process or with the provision of services, price will give landowners the incentive to sell. That is, the higher the price, the more they are willing to sell. In other words we again revisit the law of supply (see Chapter 1). Likewise, if we were considering agricultural land redesignated as potential development land, the farmer, if given the choice, is likely to sell off marginal land first before selling prime arable or grazing land. Moreover, the illogicality of the perfectly inelastic supply curve is that it implies that landowners would be as willing to sell land at the low price of P_1, just as much as at the higher price of P_2, as the same quantity (Q^*) is traded at both these prices (see Figure 4.3).

Admittedly though, the supply of land in many circumstances, especially at the local level, is likely to be highly inelastic. Such inelasticity comes about as the number of landowners currently willing to sell (assuming no compulsory purchase orders have been served), general planning constraints, and tight green belt policy will all play their part in limiting the amount of land available for development at any given time. However, normally, even in a highly developed urban environment, some land will flow on to the market as sites are used more intensively, old buildings are demolished, e.g. due to functional obsolescence, and there is a general

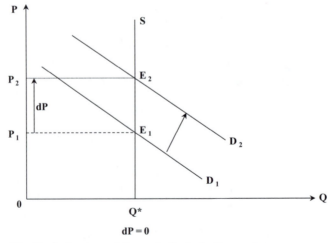

Figure 4.3   *The illogical nature of the perfectly inelastic supply curve*

expansion of urban areas. Moreover, as suggested above, landowners are likely to behave like any other entrepreneur and would not necessarily sell simply because they had a vacant plot of development land. It would be unwise to sell land in the depths of a severe economic recession when such land could become prime commercial or industrial land commanding a high price during an eventual upturn in the economy.

Another important issue to consider is that the demand for building land is a derived demand. That is, construction firms and developers, unless they are extremely eccentric, do not just demand land for the sake of it. They demand land because they can build upon it, either straightaway or at some future date, and hopefully sell the completed development at an acceptable profit. Thus, the price that they are willing to offer the landowner will normally depend upon the state of the market for completed buildings. However, in the case of the speculative builder, or for a development that is not pre-sold and will take some time to complete, the anticipated future value of buildings will need to be forecast.

For example, take the case of house builders. During a boom in the housing market when house prices increase rapidly, house builders will realize that as demand is high, they could build and sell more houses, and at higher prices, which in turn should enhance their profitability (see Chapter 8 for information on costs and revenues). Thus, if building firms feel that the boom will be sufficiently sustained, they will look around for more available building land. Consequently, the demand for building land increases, and land prices rise. Therefore, the increase in land prices has been derived from the original rise in house prices. This is diagrammatically shown in Figure 4.4. Here, it can be seen that as house prices rise from P_1 to P_2, due to a rise in housing demand from D_1 to D_2, $Q_{H1}Q_{H2}$ more houses are built upon $Q_{L1}Q_{L2}$ hectares of land. To tempt landowners to part with this land ($Q_{L1}Q_{L2}$) land prices are bid up from P_1 to P_2.

The price that a builder will be willing to offer the landowner is frequently referred to as the residual. That is, what is left over after profits and all

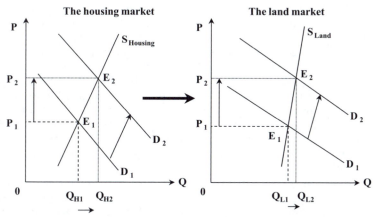

Figure 4.4 *The derived demand for land*

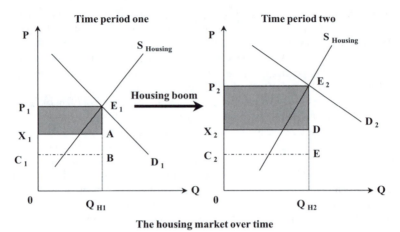

Figure 4.5 *The origin of the 'residual'*

construction costs have been taken into account. Thus, if house prices increase, higher profits may be taken, and indeed there may well be upwards pressure on building costs as more labour and building materials are demanded, but the amount of money left over to bid for land is also likely to be greater. Conversely, if house prices were to fall, less money would be left over for the payment of the required land.

Continuing with the example of escalating house prices, and with reference to Figure 4.5, it can be seen, by viewing the graphs from left to right, that the demand for housing has risen from time period one to time period two. Such a price increase has encouraged new development so that the number of new houses being built has increased from Q_{H1} to Q_{H2}. The overall revenue received by the builder has increased from $0P_1E_1Q_{H1}$ to $0P_2E_2Q_{H2}$. Therefore, even if the builder's profit margin has increased, and costs have risen, it is likely that more will be left over for bidding for building land. That is, the excess left over is now greater. Diagrammatically this excess, or residual, can be shown by area $X_2P_2E_2D$, in time period two, being greater than area $X_1P_1E_1A$, in time period one. The areas OC_1BQ_{H1} and OC_2EQ_{H2} represent the builder's non-land costs, and areas C_1X_1AB and C_2X_2DE represent the builder's profits in each instance. Thus, quadrangles $X_1P_1E_1A$ and $X_2P_2E_2D$ are the areas under the demand curve providing revenue for the purchase of land. In each instance these shaded areas simply denote the amount of money at the builder's disposal to bid for building land after the subtraction of other costs and profit. This can be seen in Figure 4.6 which shows the residual (shaded) areas transferred from Figure 4.5. Note that the sizes of the transferred areas have been exaggerated for the purpose of clarity.

Variations to this theory can be examined, such as the fact that the land used for a current development may be acquired from a previously obtained land bank. In such a case, the residual simply represents the amount of money available to purchase the next area of land for the next development.

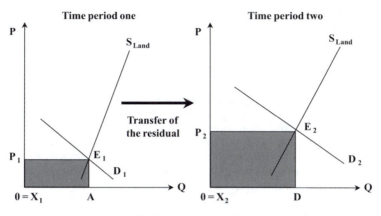

Figure 4.6 *The demand for development land*

The market for construction materials and components

The decision by the builder to use certain building materials or components will depend upon a wide range of considerations such as:

- The range of materials available on the market to chose from. The variety of materials available will largely depend upon the number of firms in the materials supply sector, the degree of competition between these firms, and the level of import penetration of such goods.
- Building standards. The materials used must at least conform to existing policies and guidelines on structural soundness, and health and safety. Components that are known to be structurally unsound, or materials that are potentially hazardous to health must not be used, and therefore act as a logical constraint upon the decision-making process.
- The quality of materials. Differing levels in quality are available for the majority of construction components and materials. A builder could select lower quality materials as long as they conformed to existing standards in an effort to save on costs. Indeed many components such as wall ties, plumbing and so on cannot be seen by the client, and therefore some aspects of quality may only be noticed by their visibility to the building's users. Alternatively, the builder could select high quality, high cost fittings such as bathroom suites and kitchen units (in the case of housing) in the hope of advertising these features as a selling point of the house. The quality of materials could also depend upon the type of building under construction, whether it is industrial or residential, and the estimated economic life of the building would also be an important consideration. On the latter point it would be unwise to fit out a building at a high cost if it was expected that it would be functionally obsolete within ten years, for example, and then demolished to make way for a new building (see Chapter 5).

- The cheapness of materials. Obviously the cost of materials is often married to their quality and therefore the discussion directly above is relevant. However, cost and quality are not always the same. For example, during a recession, because of falling demand, many construction components may fall in price although their quality remains the same. Furthermore, relative costs may change as new production techniques are found, and new producers arrive on the market to compete with existing suppliers. An example of the latter case is cheap imports from abroad that undercut existing domestic suppliers.
- Consumer demand. With regard to visual components and materials, builders would be wise to ensure that they keep abreast of current consumer tastes as the installation of popular items can be used as a selling point of the building in question. Alternatively, builders may allow their clients to select a variety of fittings according to their own specific requirements and tastes. For example, many house builders allow the purchaser to select such items as tiles, bathroom suites and kitchen units.

All of these variables, and others, will affect the demand and supply of any material or component. It would be a laborious exercise to go through each of the above points in turn in any great depth but they are all worthy of consideration. The following example examines the case of window frames. In particular, three similar products are discussed as they can be seen as near substitutes in that they perform the same task: namely plastic window frames, wooden window frames and aluminium window frames.

Generally, in periods of high construction activity more buildings are constructed and thus more window frames are required. Moreover, if the building boom is, as will probably be the case, due to an economic recovery, more people are also likely to be able to afford new replacement windows for existing buildings. Window replacement may be seen as desirable if the existing windows are in a poor state of repair, or if the new windows have superior qualities such as enhanced energy efficiency, or better draught and noise exclusion. Thus, the demand for all types of window frame is likely to increase, and depending upon their elasticity of supply, this will tend to have an inflationary effect upon their price.

For example, if there were only a few firms in any area producing plastic window frames, they may be unable to cope, in the short run, with any dramatic increase in demand. As such, the only way that they could ration the available supply would be via an increase in price as depicted in Figure 4.7, where prices rise from P_1 to P_2 with only a very limited increase in production from Q_1 to Q_2. However, if imports could be found from other regions or countries, the supply curve would become less inelastic and therefore demand would be spread over more firms and there would be less of an impact upon the price as seen in Figure 4.8. Indeed price competition is likely to exist between these firms in any case. Alternatively though, an increase in demand could be diverted into a substitute good such as aluminium or wooden window frames, if insufficient new supplies of the original good could not be found.

Another possibility is that one product, for example aluminium window frames, could become unfashionable in the public eye leading to a rise in

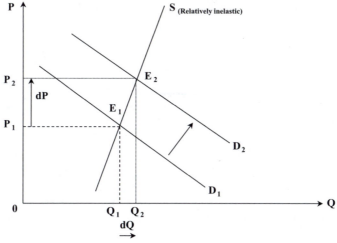

Figure 4.7   *Increasing demand for plastic window frames – the case of limited supply*

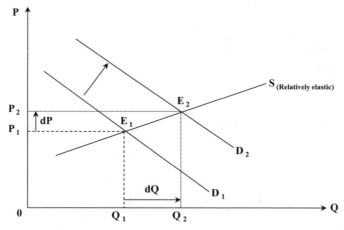

Figure 4.8   *Increasing demand for plastic windows frames – the case of extensive supply*

demand for alternatives. This outcome is likely to create a drop in demand, and thus price, for aluminium window frames as seen in the left-hand segment of Figure 4.9. The overall market result will depend upon many factors such as the available supply of each product, the degree of substitutability between the goods, and the potential existence of near substitutes such as plastic or wooden window frames. Moreover, for new buildings the limit to such substitution depends upon whether the consumer is willing to pay a higher price to cover the cost of having an alternative material incorporated into the building. If they are not, a stage could be reached where plastic or wooden frames became too expensive,

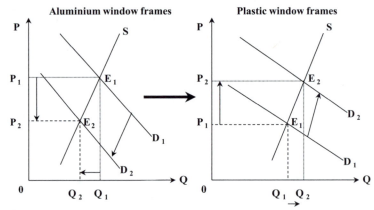

Figure 4.9   *Falling demand for one product leading to increasing demand for another related product*

due to their rising popularity and demand, and thus either aluminium window frames would be installed, or cheaper, lower quality window frames would be sought. This movement between the original good and its substitute is illustrated in Figure 4.9. It should also be remembered that the degree of substitutability can be measured using the concept of cross price elasticity (see Chapter 2).

The market for building labour

Although the general behavioural aspects of labour are similar whether one is considering surveyors, or site labour such as electricians, carpenters or bricklayers, it would be naïve to treat labour under an all-encompassing heading in an attempt to discuss the behaviour of all labour within the economy. This is because different circumstances will affect different trades and professions over time, and as such, it is best to separate them along these lines before initiating any comprehensive or meaningful study or analysis. Therefore, in this section the text concentrates upon an example of one trade, that of bricklayers, and how a variety of factors may affect their employment. In turn the potential impact of these conditions may be seen upon other related issues such as site labour productivity, the design of buildings, and the materials used.

As a starting point, imagine the overall market for bricklayers. As with any other market, there will be a demand for bricklayers (a derived demand in this instance) and a supply of people willing to follow this trade. In many countries houses are built in a 'traditional' manner with the outer walls constructed of brick. With such construction one would expect the demand for bricklayers to be highly inelastic in the short run as shown in Figure 4.10 by the demand curve $DL_{1(SR)}$. Such inelasticity is likely because even if wage

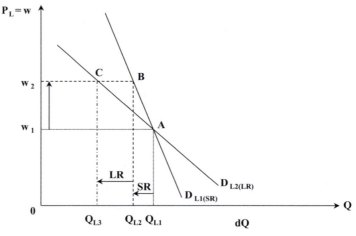

Figure 4.10 *The demand for bricklayers over time*

rates, or the amount paid to a bricklaying team per house built, were to be increased dramatically, building firms, at least in the short run, would have to employ roughly the same number of bricklayers to construct houses in the normal manner. Such a rise in payments made to labour could come about as a result of effective trade union action or if government unilaterally imposed minimum wage legislation. Obviously, in the long run, the demand curve is likely to become more elastic as some bricklayers, or bricklaying teams, could be laid off as building firms reacted to the problem of rising wages in a variety of labour substituting ways as shown below.

- Firms could insist upon higher productivity rates from existing bricklayers or bricklaying teams.
- Firms could assist such productivity enhancements by ensuring that any potential supply constraints that hampered the bricklayers from working as quickly as required, were reduced or eliminated. Improvements could be along the lines of:

  - Reducing time delays in getting materials to the workers.
  - Reducing site wastage so as to ensure that adequate materials are available when required.
  - Ensuring that production is organized efficiently so that the time a bricklayer needs to spend on a house is kept to a minimum. In other words, the site manager should try to arrange the sequence of construction in such a way that a bricklayer does not have to return to a house to redo part of the job that has been damaged or altered by another subcontractor, for example.

- Firms could simplify the design of their houses so that less bricklaying, or at least less complicated, time-consuming bricklaying, is required. For example, at the extreme, a simple 'box-shaped' house will use fewer bricks than a house of a more intricate design and layout. However, firms would have to be aware that such cost-cutting actions could also reduce the price of the house and thus revenue received, as there may be less demand for such 'utilitarian' housing units.

- Changing, or adapting, the method of construction so that less on-site bricklaying is required. The use of more prefabricated materials and units, or using alternative materials such as more glazing, for instance, could achieve this.

Referring again to Figure 4.10 it can be seen that if, in the short run, wage rates increased from w_1 to w_2, firms could not simply lay off workers as governed by the demand curve $D_{L1 (SR)}$. However, as discussed above, in the long run firms could alter their methods of production, and thus the demand curve would become more elastic. Therefore, with the initial demand curve, wage increases would cause the laying off of only $Q_{L1}Q_{L2}$ bricklayers, whereas, in the long run, as demand moves to $D_{L2 (LR)}$, a further $Q_{L2}Q_{L3}$ bricklayers could also be made redundant.

On the supply side (see Figure 4.11) the supply curve represents how many people at any given time, wish to be, or are qualified to be, bricklayers. Some of these people (Q_{L1}) will be willing to work at a relatively low wage rate such as w_1, whereas others (Q_{L2}) will only be able to supply their labour at relatively high wage levels such as w_2. The diagram suggests that nobody would work for less than w_0, as such a wage would be deemed too low to attract people into this particular trade. Presumably at a wage that low people would be better off seeking alternative employment, or receiving state unemployment benefits (if available).

Bringing the two concepts of demand and supply together one arrives at the market for bricklayers as shown in Figure 4.12. Here, the market establishes a market wage of w_m and Q_m bricklayers would be employed. As with all markets, this outcome is likely to vary at the local level

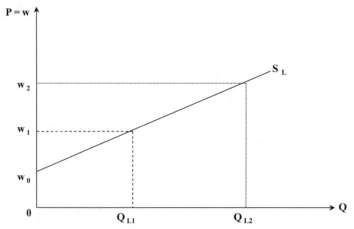

Figure 4.11   *The supply of bricklayers*

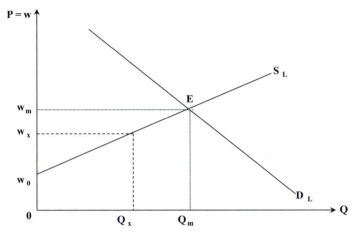

Figure 4.12 *The market for bricklayers*

depending upon specific conditions of supply and demand. For example, there may be a shortage of bricklayers in an area where there is a high level of construction activity. Such a situation is likely to drive wage levels to a higher level as demand is high and supply is constrained. In Figure 4.12 the area $0w_0EQ_m$ represents the transfer earnings of bricklayers, and the area w_0w_mE, their economic rent. Transfer earnings are the payments necessary to keep a particular factor of production in its present use.

For example, if the bricklayers represented by Q_x did not receive a wage of at least wx they would elect to work (or transfer) elsewhere, or become voluntarily redundant as presumably they would be better off working in another trade or living off state benefits (if available). Thus, transfer earnings represent the opportunity cost of working in this particular trade as the worker could earn this amount by working elsewhere. Economic rent, on the other hand, is an added bonus, a payment over and above that necessary to keep the worker on (as a bricklayer in this case). Economic rent only comes about because of the fact that the high level of demand has attracted people with high transfer earnings, such as those workers represented by Q_m on the diagram. Note that worker Q_m has no economic rent whereas workers (who may be less versatile on the job market) such as Q_x have a significant economic rent. Workers represented by Q_m may also be skilled as plumbers, for example, and as such they will only be willing to be bricklayers if the returns they receive from this are greater than those received from plumbing.

It must again be stressed that this model is a dynamic, not a static, concept. That is, both demand and supply could, and will, change over time. For example, imagine the house-building industry going into recession, say due to a wider economic recession caused by high interest rates. In such a situation it would be expected that building firms will react to depressed demand by building less houses, and therefore less bricklayers will be demanded. Thus, wages are likely to drop, say from w_1 to w_2 in Figure 4.13,

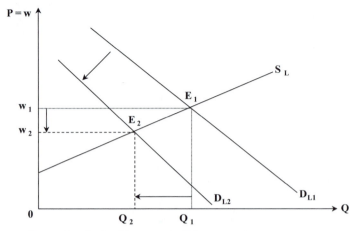

Figure 4.13   *The market for bricklayers: a building recession*

as the demand for bricklayers drops from D_1 to D_2. Thus, a new market equilibrium is achieved at a lower level with Q_1Q_2 bricklayers being laid off. The total fall in the demand for labour may be reduced as firms could react to a recession by changing the mix of their developments by building different types of houses. That is, they could find out which houses are least affected by the decline in people's purchasing power. For example, 'executive style', large houses may be harder hit than the 'first-time buyer' market and therefore builders will seek planning permission to build smaller units. Thus, the total impact on the employment market may initially be quite small.

Moreover, another consideration is that developers could try to protect their situation in such a depressed market by selling houses of a cheaper form of construction and finish so as to retain some profit as house prices decline. Conversely, however, in such situations some firms have gambled by building houses of higher specification and cost so as to find a particular niche in the market. Both of these reactions to a recession may help to protect the prospects of the bricklayer. Obviously, the above analysis could be reversed if one were looking at a housing boom. Increases in construction activity lead to rapidly enhanced demands for site labour. If such labour is in relatively short (inelastic) supply the rise in demand can lead to a rapid escalation in site labourers' wages.

Imagine the case where there is a town experiencing a housing boom, yet there is currently only a given number of bricklaying teams available in that town (say Q^* in Figure 4.14). This situation could be represented by a completely inelastic supply curve for bricklayers in the very short run as shown by $S_{(SR)}$. Therefore, as demand increased from D_1 to D_2 existing teams could work harder and earn overtime payments equal to w_1w_2. Such high wage rates, or returns, are likely to attract others to become bricklayers in the area, whether they become trained in the trade, or they travel in to work from other geographical locations. If additions to the supply of

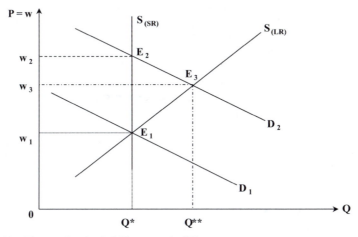

Figure 4.14 *The market for bricklayers: a building recovery*

bricklayers did come forward in the long run there would be increased levels of competition from other bricklayers entering the market giving a new supply curve of $S_{(LR)}$. Building firms could now choose from a greater pool of labour and wages would be driven down to w_3. The end result would then be that less overtime would be done, and Q^*Q^{**} more people would be employed.

Entrepreneurship and individual effort

Many individuals contribute to the value of a property and the ease of selling it. For new and successful developments most credit should go to the entrepreneur who initially envisaged the feasibility of the project and the opportunities that stem from it. However, credit also needs to go to the architect that can produce a design relevant and attractive to the market. In addition there is the skill of those involved in disposing of the development by securing a deal that will sell the property at the price desired by the owner or the developer.

In terms of existing buildings that are brought back on to the market for resale the individual can do much to enhance, or at least maintain, the value of the property. In the case of housing, for example, imagine two highly similar properties in the same area being offered for sale. The careful owner that keeps his or her property in good order will have an advantage when selling over an owner who has let his or her property fall into relative disrepair. In other words, the value and price of one building could be greater than the other due to the condition in which it has been kept. Thus, the value of the former building has been enhanced, as it is a structure that is both visually attractive and has fully functioning facilities. As a consequence its price is highly likely to reflect this and indeed the property will probably be sold in a short period of time. The purchaser of such a building is likely to be willing to pay a higher price as they are secure in the knowledge that the plumbing and heating, for example, are all in working order and are not going to lead to costs in terms of both time and money.

Such a process can be further advanced by effort in terms of home improvements and decorating as long as these are carried out in a tasteful manner that accord to current fashion. Some successful sales are attributable to the redecoration of a property immediately prior to it being placed on the market.

Moreover, the individual can encourage a successful sale by giving the impression of greater space by creating the illusion that the dwelling is larger that it really is. House-building firms selling new homes invariably remove the doors from the show home so as to create a more spacious feeling. The removal of excessive ornamental clutter is felt to have a similar effect, as is the avoidance of the use of large and bulky furniture. Indeed it has been known for firms to use scaled-down objects such as three-quarter sized beds, chairs and sofas so as to create this effect. The perception of space is also influenced by a careful use of colour. In its simplest form light colours tend to make rooms look larger than darker ones.

The individual can also work to create an impressive view of the property by careful and effective advertising. In this arena, colour photographs or artists' impressions may attract more people to an advert placed in a newspaper rather than those that are in black and white. In fact the mere naming of the property could attract clientele. For example names can conjure up images of homeliness or grandeur depending upon the market being aimed at. Promoting a property normally goes further than advertising. The talent of the site sales representative or the estate agent is obviously crucial in being able to obtain a sale at the desired price. The whole attitude of such people, even the way that they are dressed, could be a leading factor in the decision-making process of the purchaser.

Inevitably, therefore, individual effort can influence the sale and value of any building. Although, by way of example, the above analysis was confined to residential property the points expressed can be translated into the actions that are required to satisfactorily dispose of commercial property as well.

Despite the importance of individual effort, far greater influences are received from exogenous forces beyond the control of those directly involved in the project. For example, an enhancement in the price of property created by external factors or the actions of a third party is relatively common.

First, imagine a landowner who owns a small farm on the periphery of an urban area. In addition, assume that a decline in the fortunes of the agricultural sector generally has led to a decline in the value of agricultural land. However, if planners extended the boundaries of the nearby settlement and reclassified the current farmland to potential development land the demand for the property is likely to rise dramatically. Given the relative inelasticity of supply associated with land the price of the land is likely to follow suit and rise sharply. Such an increase in the demand for land can be seen in the right-hand side of Figure 4.4. Therefore, the price of the original farmland has been greatly increased due to the actions of the planners rather than an entrepreneurial decision by the farmer. However, credit could be given to the farmer as it could be argued that he or she might have had the vision to hold on to the land in the knowledge that it would eventually attract the status of building land. In the same way development companies could purchase agricultural land as a speculative investment. With such a scenario the company could lease the farm in the short run as a going agricultural concern in the hope that its classification will eventually change its potential use. However, some have argued that any windfall gain to the landowner as a result of planning decisions should be taxed as it has not been chiefly due to skill or effort.

The impact of planning permission granted for nearby or adjacent land can also have far-reaching consequences. Imagine a housing estate that is presently situated next to open countryside. Despite the picturesque value of open space, greenery and its possible amenity value, the price of the existing houses may rise if a complementary development were to occur on this rural land. For example, a retail centre providing goods and services, or a business park creating an opening for local employment, could increase the attractiveness of these houses for people wishing to locate in the area. If this were the case, demand and therefore price would increase. Of course one can easily envisage the opposite scenario taking place. The desirability of the estate could plummet, causing a crash in its house prices, if a conflicting land use were placed on the nearby land. Such conflicting land uses could be the digging of a quarry, the introduction of heavy industry, or the opening up of a civil waste tip, for example.

The price of the buildings, whether they be residential or commercial, could drop if it came to light that the land on which they stood was contaminated from previous industrial use or the dumping of waste. Here, not only is one witnessing the actions of others affecting the price of land, but the actions of a third party from an earlier time period. Even if the land is not too heavily contaminated and is not considered to be a risk to health, the mere awareness of its potential threat could devastate the selling price of the property. Highlighting the issue via the production of a register of contaminated land would formerly bring this problem to the attention of any purchaser, and uncontaminated land near problem areas may be similarly avoided by players in the market. However, for occupants who do not wish to sell their property it is true to say that although the sale price of their buildings may have fallen, its use value will remain unchanged.

Property prices can also be influenced by the whims of fashion. For example, an area of housing or retailing may command high prices and rents when they are said to be in the fashionable area of town. However, either the regeneration of another area or the development of a new one may alter the original attractiveness of the location. A change in shopping habits from shopping in the city centre to out-of-town shopping, perhaps primarily induced by the mass ownership of the private car, would obviously cause a restructuring of a town's property price and rent contours (see Chapter 5).

A change in the fortunes of industry can have a dramatic impact upon local property prices. A town may be dominated by an industry that produces goods or services that become high in demand. The success of the industry is likely to lead to more employment and higher incomes in the area, which in turn would stimulate demand and subsequently prices for all forms of property. Conversely, industrial decline caused by international competition or a collapse of the market, for example, could lead to widespread unemployment, a reduction in local incomes and property blight.

General government policy, especially at the macro-economic level, whether it is fiscal or monetary policy (see Chapter 12), could have a radical impact upon property prices. For example, a significant and sustained lowering of the interest rate, to choose just one variable of demand management, could raise the demand for goods and services and thus enhance the markets for office and retail space. More directly, a lowering of the cost of borrowing would create a more positive climate for the residential sector through cheaper mortgage finance.

Finally, although this list is by no means exhaustive, the price of a building can obviously be adversely affected by the discovery of a serious structural defect or dangerous material such as asbestos.

To conclude, although both the entrepreneur and user of the property can influence values they are dramatically affected by external factors largely beyond their control.

Productivity in the construction industry

Productivity is essentially the measure of output that is produced by various inputs of capital and labour in the productive process. As with all industries, maintaining a high level of productivity in the construction industry is important in order to assure its long run profitability via client satisfaction and speedier debt repayments, and its international competitiveness. However, when attempting to measure productivity a number of difficulties need to be addressed:

- There are a large number of statistical sources, some of which use different definitions of productivity. For example, some only measure output per actual employee of the construction firm whereas others look at the output of the whole firm including the contribution made by subcontractors. The difficulty here is that the former figure will overestimate the contribution to total output of the firm by each employee as it does not include the contribution made by others hired by that firm in order to achieve its output.
- Some data sources do not take into account the erosive power of inflation. Therefore, when output is expressed in terms of monetary profit, or turnover, what appears to be constant growth over time may only be reflecting the declining value of money itself. For example, if a small building firm built two houses a year at a cost of £100,000, and the following year it only built one house, yet construction costs had doubled, it would appear that its productivity, measured in terms of materials used in this instance, had remained the same at £100,000, yet it had in fact

halved. Therefore, it is best to use data that are given in real terms as such data have taken inflation into account.

- Financial figures can also be misleading when comparing statistics between different countries as variations in the exchange rate can radically alter the overall view. For example, if a worker in a country is seen to produce an output of $10,000 per year and then the currency of that country is devalued in relation to other currencies, productivity will be seen to rise in money terms, say to $12,000, yet their physical output could remain unaltered.
- Different sets of data may contain different definitions of the construction industry itself. One obviously needs to be clear about the exact item that is being measured, otherwise misleading additions or omissions could occur. This problem is especially noticeable when examining international comparisons of productivity, and looking back at time series data in countries where the definitions of industries have changed over time.
- Variations in figures relating to productivity may occur simply because researchers have used different sample sizes. As such one must always take care that one's study sample is statistically significant. Regional variations in construction productivity may also have an impact upon the overall figure.
- Evaluations of productivity in the public sector of the construction industry frequently reveal far lower figures than those for the private sector, thus giving 'ammunition' to those who point to public sector inefficiencies. However, much of the difference between the two sectors could simply be explained by the fact that the public sector often has a greater load of repair and maintenance work rather than new build and therefore additions to output appear low.
- It is felt by many that official estimates of the construction industry's productivity may be artificially low as firms fail to reveal the true picture of their total output. The reasoning behind this statement is that because of the large number of very small firms and the growth of self-employment in the industry, many do not make full and accurate returns to the taxation authorities. For example, some builders are well known for 'moonlighting'. In other words they undertake work in exchange for cash payments so that taxes can be avoided, and it is in this manner that much small building work may not appear on official records of construction activity.
- In the northern hemisphere much building site labour is employed on a seasonal basis as most construction work occurs in the warmer parts of the year. Due to the seasonal nature of such employment, output per worker looks significantly lower than for most other industries that are engaged in production all year round in the shelter of a factory or an office. Adding to this problem is the fact that construction contracts are discontinuous and a firm may not be able to guarantee involvement in a new development as soon as another is completed.

So far this analysis has implied a concentration on the productivity of site labour. However, in assessing the efficiency of any building firm one should also look at the productivity of the other factors of production employed by the firm such as capital and entrepreneurship (management). As touched upon above, the importance of enhancing productivity in the construction industry is paramount for a number of reasons.

- Higher productivity can lower a firm's costs, as, for example, it requires less labour to complete the same task. In the same way the firm will need to hire capital and subcontractors for less time, and in addition there is only a need to borrow development finance for shorter periods. Higher productivity can also increase the revenues of the building firm as there is evidence to support the view that clients are willing to pay more for a job that is done quickly and efficiently as they can then use the building for raising revenue themselves. As revenues are enhanced and costs decline, the profitability of the building firm is increased.
- Such improved productivity maintains the competitiveness of firms against both domestic and foreign rivals when tendering for work at both home and abroad.
- It is argued that improvements in productivity help to keep construction costs down and such savings can be passed on to the client in the form of reasonably priced buildings. If the cost of completed buildings were to increase too rapidly when compared with other inflation in the economy there would be a risk that people would 'make do' with older, existing buildings via refurbishment and repair and maintenance, rather than commissioning new build. A commonly cited reason for not ordering a new building is that clients are worried that development costs will escalate during the actual construction phase making the project far more expensive than originally anticipated in their investment appraisal calculations.
- Some researchers have suggested that high productivity in the construction industry is imperative for correspondingly high productivity in other areas of the economy. For example, for a manufacturer, or a retailer, to remain competitive the buildings that they require must be completed efficiently and on time. In fact evidence shows that productivity is one of the best indicators of overall economic success as those countries that have high increases in output per person have experienced the greatest economic growth.

As suggested above there are a variety of reasons why construction productivity is difficult to measure and may seem lower than it actually is. However, despite these statistical defences of the performance of the building industry there are also a number of physical characteristics of the construction process that will impair production efficiency.

- There are no long, standardized production runs as contracts are varied. Each and every building is on a unique site with its own special characteristics and therefore the production process has to be reorganized again and again. Moreover, even if all building components are completely factory produced, they still have to be assembled on site. Although there is much assembly of prefabricated units, a great deal of preparatory groundwork is still required as seen below.

- The site needs to be cleared before construction can commence. This may necessitate the demolition and removal of an existing structure.
- Work needs to be carried out on the foundations of the building. Such work is complicated by the fact that significant variations in site conditions can occur with different rock structures, terrain, and the presence of mining works.
- The connection of services such as gas, electricity, sewage disposal and water needs to be organized.
- Access problems to and from the site need to be dealt with. Often changes to the urban road system need to be organized in order to accommodate the new development and its likely use.
- Even with the best initial surveys and plans unforeseen construction problems can arise on a case-by-case basis requiring one-off 'on the spot' decisions being made.
- The individuality of each building design is largely at odds with the standardized, mass-produced, nature of goods and services in modern industrial societies. There is also a geographical separation of sites that causes breaks in the flow of production, which subsequently necessitates that both labour and capital need to be mobile.
- There is discontinuous demand, and labour is not necessarily employed all year. In other words, a construction firm cannot guarantee work on a new site when its current development is complete.
- Either the client or the design team often delays the completion of a building because of frequent changes in specification after building work has commenced. Such changes are often very difficult and time-consuming to achieve. Moreover, any major changes may have to be agreed upon by the planning authorities, which again adds time to the development process.
- Pressure groups can also delay construction due to particular objections to developments. For example, environmentalists may object to the nature of a building in a certain area. In a similar way those who wish to conserve the country's historical past may object to a building that may encroach on a historic site or remains.
- The planning process itself if often cited as a slow and lengthy one thus adding time between a project's inception and its completion.
- The construction of buildings is frequently delayed because of slow decision-making and misunderstandings between the design side, the building firm, subcontractors, and the client.
- In the case of complex engineering structures and unique designs, that do not use standard construction techniques or materials, the development process can be delayed as unexpected and unknown difficulties are encountered.
- The types of building that the public or clients desire often do not lend themselves to rapid construction. There tends to be a preference for buildings, especially in the residential sector, to be built in a 'traditional manner' rather than using more modern construction techniques such as the increased use of prefabricated sections.
- In times of prosperity in the construction industry, so much building work is being undertaken that the supply of crucial skilled labour can

become scarce. In such situations projects have been delayed because key tradesmen, such as plumbers and electricians, cannot come to the site immediately as they are engaged elsewhere. These delays can have a cumulative 'knock on' effect as other tradesmen, say plasterers for example, cannot easily perform their tasks until the electricians and plumbers have completed their work.

- Some researchers have claimed that a further reason for low construction productivity is the poor labour relations and working conditions found on most building sites. Many sites are characterized by a lack of clean, dry, relaxation areas, with which most employees in manufacturing would expect to be provided.
- There is also evidence to suggest that poor management, especially at site level, is another reason for the slow completion of developments. It is a noticeable fact that some firms manage to organize, initiate, and complete production at a far faster rate than others.

As has been demonstrated, there are a large number of reasons why construction productivity may be lower than productivity levels found elsewhere in the economy. Moreover, some countries may experience even lower productivity levels than others due to the following factors.

- Building standards or regulations may be higher, or more detailed, in one country than in another. Higher standards are likely to necessitate more careful and detailed building using better materials and are therefore likely to add time to the construction process.
- Poor weather, for example heavy frost, is a frequently cited reason for delays in the construction process, so much so that building contracts often have clauses in them to allow for delays caused by unfavourable weather conditions.
- The planning system in some countries is more efficient and less bureaucratic than in others, and therefore delays at the early stage of the project's life can vary.
- The demands from the public and clients vary from country to country. Increased use of prefabrication may be quite acceptable in one country, whereas in another country, slower, more traditional construction methods will be insisted upon.

There are numerous problems to be overcome with respect to construction productivity in order to ensure that the industry is in a suitable state to meet the increasing demands of the modern economy and international competitiveness. As such there now follow some suggestions on how the development process could become more efficient.

- In many countries it is common to see contracts negotiated between the construction firm and the client whereby it is agreed that any ongoing increases in building costs are largely reimbursed by the client. In this way the building firm will not suffer any erosion of its profits if costs increase during the actual period of construction. (However, as costs do tend to increase over time due to general inflationary pressures in the economy, it can be argued that if the building firm was forced to sign a fixed price contract, whereby increases in costs would not be reimbursed by the client, the building firm would have more of an incentive to complete the development as quickly as possible before cost increases reduced its profit margin on that project.)

- Financial penalty clauses could be written into the contract so that if the construction process was late at any agreed stages, especially at completion, the building firm would have to pay a fine to the client for failing to reach its target objective. In this way one would expect the building firm to have an incentive to enhance its organizational efficiency. In order to implement such a policy the client should be provided with regular progress reports.

- Efforts could be made to improve the performance of subcontractors and materials suppliers as it is often failure on behalf of these groups that lead to construction delays. Perhaps the construction firms could impose fines on these parties if they failed to reach their promised deadlines.

- As clients cause many delays themselves, they must be encouraged to consider their requirements carefully before the construction process commences so as to limit delays created by subsequent variation orders. Moreover, there is often a need for clients to provide more detailed information about their specific requirements so as to ensure that they are catered for.

Innovation in the construction industry

Because of the apparent low level of productivity in the construction industry many perceive it as an industry that is characterized by backwardness and one that is generally very old-fashioned. A backward industry can be defined as one that uses out-of-date working practices and techniques, so that it produces a product in an antiquated manner. If an industry is in this situation it is felt that it will not achieve either full efficiency or profitability. Therefore, there is pressure for construction firms to modernize their operating practices in order to meet the requirements of the modern economy and to remain internationally competitive.

However, it must be noted that striving for efficiency and profitability may not be in the complete interests of society. Buildings could perhaps be built at a faster rate, however they may become more basic, lacking in design intricacy, and become heavily reliant on prefabricated components. Such a situation may lead to the creation of a very utilitarian built environment with many similar buildings being built of cheap construction. However, it could perhaps be argued that it would be possible to reserve this form of construction for factories on industrial sites, away from the main public view. This argument specifically holds for industrial buildings that may have a very short economic life (see Chapter 5) and therefore it may be extravagant to build such buildings in anything except the most basic way. Despite these fears, and the increasing movement for many new buildings to be built in a more intricate way so as to mimic the architecture of the past, there are still arguments, as seen directly above, for increasing the level of innovation in construction.

Innovation should provide ideas about how to improve the efficiency of working practices and the techniques of production so that a modern product is supplied on to the market. It should be appreciated that such techniques can be incorporated into the construction process, yet at the same time giving the completed building a traditional appearance. For example, a steel-framed structure with prefabricated walls could still have a traditionally built and designed façade, or at least one that gave the appearance of being old-fashioned. In fact, in many inner city renewal programmes, older buildings have been demolished except for their frontages. These frontages have been retained and a newer structure has been built on to them at the rear. This gives the dual advantage of a modern purpose-built building that is aesthetically pleasing and conforms to the traditional and historic view of its location.

When discussing potential efficiency improvements in an international context one must be careful not to make global recommendations, as it must be recognized that the types of buildings constructed often reflect the relative costs of both techniques of production and materials used. For example, in a country where both bricks and manual site labour are expensive, there is more of an incentive to use alternative materials such as timber which lends itself more readily to prefabrication and subsequently reduces the need for large numbers of site labourers.

Advances in technology can be categorized under a variety of different classifications:

● Modern materials and components technology.
● Enhancing the use of capital.
● The use of 'modern management' techniques.

In the case of modern materials and components technology it is felt that increasing the amount of off-site, factory based, production should enhance productivity. In other words, using more prefabricated components should facilitate a speedier construction time, although the process of assembly would still need to take place on location. In modern times it can be seen that the building process has increasingly become an assembly operation of previously manufactured components, and that such a growth in component technology is likely to take place as long as components are economically transportable from their place of manufacture to building sites. It should be remembered that advances in technology could make what seems impossible now quite commonplace in the future. For example, at one time even bricks were manufactured on site!

In terms of the degree of capital input into the industry, many argue that a low capital to labour ratio makes the construction industry automatically seem backward, as most modern industries are highly capital intensive. The capital to labour ratio is simply a measure of how much capital is available to each worker in an industry. For example, a low capital:labour ratio would be given by the figure 1:50. This implies that there is only one unit of capital to every fifty workers. Therefore, conversely, a high capital to labour ratio of say 1:2 implies that there is a unit of capital available for every two workers in the industry. However, improving the capital to labour ratio in the construction industry is problematic and possibly not the most logical thing to do for a variety of reasons as suggested below.

- The construction industry is an assembly industry that assembles components that are manufactured by other industries. It is argued that capital intensity in the form of powered plant lends itself easily to manufacturing, yet not so easily to the assembly process. Moreover, manufacturing is often concerned with a homogeneous good whereas construction is usually involved in highly varied production.
- There is little potential for the application of a great deal of capital in the repair and maintenance sector of the construction industry. Therefore, any data on capital use in the building industry may appear misleadingly low as repair and maintenance is normally a very large sector of the overall industry.
- As implied above, the capital to labour ratio will obviously depend upon the relative costs and availability of the various factors of production such as capital and labour. For example, if site labour is more expensive in one country than in another, the capital to labour ratio is likely to be higher in that country as capital usage becomes relatively cheap.
- The construction industry is usually dominated by a large number of small firms, and such small firms will rarely have sufficient finances available to acquire much capital equipment. Moreover, the size of contract that they are involved in is unlikely to be large enough to warrant high capital usage.
- The need for the ownership of capital equipment by construction firms is not great in many countries as a well-developed hire sector tends to service their requirements. The hire sector has evolved as much plant is highly specialized and subsequently too expensive to be afforded by the small firm. Moreover, as firms will normally be involved in a number of different contracts over time they do not want capital to be tied up in a particular piece of equipment that is site or job specific. Furthermore, construction contracts are often discontinuous, and therefore firms will not wish to incur the expense of having capital equipment lying idle in the event of them not securing a contract immediately after the completion of the existing one. Finally, building firms are reluctant to purchase equipment that is easily damaged in the rough environment of a building site. Therefore, if data is examined on the use of capital in the construction process it may give an artificially low figure as the capital is owned by firms in the hire sector rather than the building firms themselves.

Alternatively, some argue that the way forward to improve productivity in construction is for the industry to take on board 'modern management' technology. The theory of modern management has its origins in the early part of the twentieth century when there was an increasing need for improvements to be made in manufacturing productivity so as to cater for the mass production necessary to meet the needs of rapidly escalating international demand. The name most frequently associated with these ideas is Frederick Winslow Taylor leading to the concept often named 'Taylorism'. The main thrust of modern management theory is that the labour force should be organized so that there is a division of labour leading to the specialization of labour. Therefore, instead of hiring workers who would perform a variety of tasks on site from bricklaying to plumbing, one should have specialists who only performed one task. In this way it is hoped that workers will become highly skilled, and therefore highly efficient, when performing their particular task. The additional motive behind this idea was that it would de-skill the workforce converting all labour's output into physical, non-mental energy that would not challenge the managerial hierarchy or its authority. The idea of increasing output via specialization was certainly not new as it had already been discussed by Adam Smith in his work *An Enquiry into the Nature and Causes of the Wealth of Nations* (1776). However, the negative aspect of such specialization is that workers can become bored by their repetitive tasks and therefore they may begin to work in a slow, despondent way paying little attention to the quality of their work.

A further aspect of the theory is that management should undertake time and motion studies in order to determine a fully productive workload for the labour force. In this way management could then attempt to control tasks by the speed of machinery, such as conveyor belts, and the succession of individual components to complete the final good.

However, it could be argued that the ideas of modern management are already incorporated into construction management. Consider the following examples:

● There is a clear division of labour into the various parts of the construction process such as plasterers, carpenters, electricians and bricklayers.
● Wages for work such as bricklaying are often paid on an incentive or bonus basis that encourages workers to complete tasks as quickly as possible. This feature highlights the need for careful site management to ensure that tasks are undertaken to a desired quality level.
● Although methods of production that use conveyor belts are not strictly possible in the construction industry as the product is stationary, one can make the labour force pass along the product line with sequential deadlines to complete their individual tasks. In this way there are great similarities in the productive process of both manufacturing and building except that in the case of construction, labour passes the product rather than the other way around.

Such ideas to improve productivity via advances in technology also need to be supplemented by the targeting of other issues that tend to retard productive efficiency in the construction industry. Delays in the building process have also been blamed upon the following influences:

- The climate is frequently said to be the culprit for the slow completion of a building. The changing of seasons can have an impact upon both the climate itself, and the number of hours of daylight. Moreover, there are daily variations in the weather, which can hinder production. For example, heavy ground frosts can make any ground work, such as the digging of foundations, virtually impossible. Therefore, unlike the enclosed conditions of the factory environment that are associated with the manufacturing industry, the construction industry is open to the natural elements, which affect standardized production and uniform working conditions. Even if advances in building technology can reduce the problems created by the weather, one still has to be concerned with related issues such as the safety of workers in poor light, for example.
- Architects producing unviable, and over-complicated ideas, or not making their designs clear enough at the outset so that remedial work needs to be carried out during the construction process.
- The planning process is often criticized for being slow and bureaucratic which can lead to delays in the commencement of construction after the initial inception of the project.
- Poor management and the lack of co-ordination of subcontractors are frequently blamed reasons for the slow completion of buildings.
- One group of subcontractors damaging the work of another group of subcontractors that again leads to the need for remedial work to correct the damage.
- New materials, or materials that have been used incorrectly, can lead to the threat of failure. Therefore, there is a need to correct the work before handing the building over to the client.

In conclusion on the topic of productivity in the construction industry the following points can be made:

- Due to economic pressures, changes in the method and type of construction are only likely to occur if such actions enhance the profitability of building firms.
- There is increasing social pressure from the general public, who form the client base of the construction industry, to construct new buildings in a traditional manner and design. Furthermore, there is additional pressure from groups such as the environmentalist movement to conserve existing buildings. However, the use of traditional techniques of construction could prevent or delay the adoption of productivity enhancing techniques of production in the construction industry.
- Many feel, though, that some innovation is necessary as building costs often increase more rapidly than the general level of inflation. Therefore, in order to maintain output, builders need to find ways of keeping prices down in order to sustain the level of demand for buildings.

● Many building firms are reluctant to use new technology, or new products, due to the degree of risk involved in using new ideas. This feature is especially prevalent in the construction industry due to the large sums of money involved in most projects coupled with the low level of financial reserves of most small building firms.

All in all a wide range of issues have been discussed so as to give a broad picture of the features affecting the factors of production that are required to assemble a development.

5 Market forces, the property investor and their role in the creation of the built environment

The private sector is the leading player in the creation of the built environment in most modern economies. Initiatives from the business community provide the main stimuli behind urban growth and change. To bring a business plan to fruition an entrepreneur may require the physical development, or redevelopment, of factories, warehouses, offices or shops depending upon the nature of the project. Indeed, even new residential estates are built to provide the developer with profit. In all cases if profit is not forthcoming development will not occur. More importantly those developments promising the highest profits are likely to be favoured. However, it must be recognized that any investment decision relating to property is influenced by controls and constraints imposed by the public sector. Public sector intervention in the decision-making process is typically in the form of planning or building regulations.

Recognizing the importance of the private sector this chapter examines a variety of common private sector investment appraisal techniques. These investment methods demonstrate how entrepreneurs make decisions concerning buildings and their location. These decisions can be at either the micro or macro level whereby the underlying objective of both is the optimization of returns from any capital outgoing. In other words, firms will wish to strive for the maximization of profit (see the discussion on the motives of business in Chapter 8).

At the micro level decision-makers would wish to utilize a suitable investment appraisal technique that enabled them to examine the possibility of making profitable changes to an existing building. For example, one may wish to investigate whether a building's current operating costs could be reduced via an investment in energy-saving features such as roof and cavity wall insulation or improved glazing. If the running costs of a building are reduced the gap between revenues received from the rent and the general costs of the building increases so that profits rise. Moreover, noticeable improvements in a property may also enable the owner of the building to increase the rent, as the building becomes a more comfortable and attractive place to work in. For example, with the installation of double-glazing not only does the building become warmer and therefore more pleasant in cold weather, but such additional glazing also tends to insulate the building from potentially annoying exterior noises such as those created by nearby traffic. Thus, in the absence of public sector interference to promote energy saving in buildings, it is likely that investment in this area will only occur if it produces a worthwhile level of profit. If buildings were more energy efficient the additional environmental benefit would be less pollution as less power generation would be required.

One may also wish to investigate whether enhanced profitability could be promoted by a change in use. For example the owner or manager of an old industrial building may find that due to the building's age it is unsuitable for most forms of modern manufacture. As a consequence it is probable that the building only attracts low paying tenants or has to be split for multiple occupancy if left in its current use category. Therefore, if planning permission and finance were to be forthcoming, the building could undergo extensive interior conversion in order to change its use. If appropriately located an old industrial building may find a new lease of life by being converted into a number of modern executive apartments for example. In many parts of the world there are numerous examples of such conversions especially in waterside industrial buildings such as those found in old dock areas. These buildings provide the executive with a residence in an imposing building of architectural interest as well as giving him or her potential access to pleasant views across water. Conversion has the advantage of conserving older buildings and structures, even if in some cases it is only their façades, rather than demolishing them and replacing them with new ones (see the debate on the economic life of buildings later in this chapter). As a result, conversion helps to preserve the architectural and historical past of the built environment.

Alternatively, profits could also be enhanced via the extensive refurbishment and modernization of an existing building. For example, an ageing office block may lack sufficient amenities with which to attract high rent paying occupants. However, if the building were to undergo extensive refurbishment whereby its facilities were improved the owners could insist upon an increase in rent so as to reflect the higher level of provision. If the existing tenants did not wish to pay an increased rent alternative higher paying tenants could be sought at the end of any current lease arrangement.

The same investment appraisal techniques can be used at the macro level. At this level the investor would be concerned with finding and using the correct investment method to make large-scale decisions about buildings. For example, at the macro level the investor would be involved in issues such as deciding whether or not to redevelop a site completely by knocking down an existing building and replacing it with a new one. Alternatively the development of a completely new site on previously non-urban land may be considered. Obviously such decisions involve much greater amounts of money than was the case at the micro level and therefore they warrant a great deal of care and precision.

Whatever the level of investment under consideration the investor needs to address some fundamental issues at the outset. Essentially it needs to be assessed whether or not:

- The investment is profitable. If the investment is not profitable it is unlikely that it will be undertaken by firms in the private sector.
- The investment is the most profitable option available. Obviously investment funds are not unlimited and therefore if there are several projects to select from one needs to attempt to ascertain which one(s) will give the greatest financial reward.

In addition to these basic investment criteria the investor also needs to ensure that the investment is affordable. Just as importantly it must be recognized that all projects carry a degree of risk. The existence of risk and uncertainty dictates that the outcome of an investment is never completely guaranteed. For example, the redevelopment of a site may incur unexpectedly high increases in building costs. A rise in costs may be due to unforeseen construction difficulties such as the discovery of archaeological remains, uncharted mine works or a build-up of methane gas in the ground from a previous landfill, for example. Similarly, anticipated lettings and rentals for the completed building may not be achieved if the economy and property market were to experience a recession that had not been forecasted. The degree to which risk and uncertainty is considered depends upon whether the investor is a risk taker or is risk averse.

Once these fundamental issues concerning the viability and profitability of a project have been assessed it needs to be ascertained which investment appraisal technique is the most appropriate for the investment in question. The choice of technique is absolutely critical, as each method is best suited for a specific investor requirement. Essentially the methods fall into two broad categories: conventional techniques and discounting techniques. Both of these categories are now discussed and compared in order to demonstrate the applicability in a variety of circumstances. In order to assist the explanation a simple hypothetical case study is provided below.

Imagine that there are two potential investment opportunities available to a property developer: project A and project B. These investments are deemed to be highly similar as they both involve the construction of buildings with identical forecasted total development costs. These costs would include monies required for the acquisition of the land, demolishing any existing structures, clearing the site, professional fees such as the salary of an architect, and erecting the new building or buildings. The similarity of these investments is compounded by the view that, at least at first sight, their returns are highly comparable. In order to simplify the analysis yet further it has been assumed that the anticipated operational project life is just four years. This assumption is for the purposes of illustration only as it is somewhat unrealistic in the case of buildings, which are normally characterized by significantly longer economic and physical lives. However, the assumption does not affect the application, accuracy or methodology of the investment appraisal techniques discussed. Furthermore, it should be noted that the anticipated net income flows generated by the letting of these developments on completion exhibit exaggerated differences and fluctuations. Again these figures have been selected to highlight various issues and will not hamper the overall conclusions gained from the analysis. Total development costs and

Table 5.1 *Hypothetical investment data*

Year	Project A	Project B	Cash flow description
0	(2500)	(2500)	Initial development costs
1	1500	1000	Net income flow, year one
2	1250	1000	Net income flow, year two
3	250	1000	Net income flow, year three
4	250	1000	Net income flow, year four
Gross	3250	4000	Gross future income flows
Net	750	1500	Net future income flows

forecasted income flows for both projects are shown in Table 5.1 where all figures are expressed in $000s.

The figures for net annual income flows would be derived by subtracting the running and management costs of the building from the receipts gained from letting the floor space. Operating costs include general expenditure on repair and maintenance and buildings insurance, for example.

Note that:

- All the figures in the table are estimates as they involve predictions of future events such as changes in the value of money due to inflation, or changes in the level of tenant demand for the projects themselves.
- The anticipated income stream from project B is uniform yet for project A it is anticipated to be high at first before dropping substantially in the future.
- In the short run the future income flows for project A are higher than those for project B.

Such a variation in potential income is obviously exaggerated as the time-scale of each project's life has been compacted considerably. However, a tendency for such fluctuations exists although on a less dramatic scale and could be explained by a variety of factors. For example forecasted net returns could be reduced for certain future dates, as the project assessor may know that the type of building being considered requires significant and costly remedial work after a certain period of time. Alternatively, it may be known that another competing project will be constructed nearby in the next couple of years. Such a development could take away the existing development's high paying tenants as they move into the more modern environment of the latest development.

Conventional investment appraisal techniques

Conventional techniques of investment appraisal are commonly used especially at the micro level. Their popularity primarily stems from the fact that they are simple and straightforward to calculate. However, they are perhaps too simplistic when the investor is dealing with larger sums of money at the macro level as they do not take into account the time value of money (see discussion on discounting techniques later in this chapter). The two conventional techniques considered by this text are the payback method and the average rate of return (ARR).

The payback method

The payback period of any project is the time that it takes for the investment to recoup its initial outlay. Based upon this the payback method of investment appraisal simply recommends projects that repay their cost quickly. The faster the repayment the better the project. Using the numerical example in Table 5.1 this method would suggest that the developer invested in project A in preference to project B as the former would recover the investment cost in less than two years, whereas the latter would take two and a half years. This preference can be expressed in notation form as:

$$PB_A > PB_B$$

One potential defect of this approach is that it just investigates the liquidity of an investment and fails to take a longer-term view. In this instance the method has ignored the fact that over the whole life of the project the net income flow received from project B is far greater than that received from project A. Indeed the net receipts from the second project are double those of the first. At first this criticism would seem so damning as to negate any further debate on the technique as a plausible measure of investment. However, there are still a number of advantages of using this method of appraisal besides its attractive simplicity. For example:

- It is a useful technique for those investors who need their money back as quickly as possible. If the value of the initial capital outlay is rapidly recouped two advantages are immediately obvious. First, investment monies can be reused to fund yet another development. Second, any development finance raised from borrowed funds can be repaid quickly. Not only does this save on interest repayments but it is also likely to boost the confidence of both shareholders and lending institutions as they witness liquidity and prompt repayment. Therefore, this method recognizes that monies received now, or in the near future, are worth more to the investor than money received at some distant date. In other words there is an opportunity cost involved in waiting for delayed returns.
- The method is useful when applied to projects that are vulnerable to change. An example of such an investment scenario is investing in the arena of rapidly changing technology. In this case projects may not be able to recover their investment costs if they become obsolete. Obsolescence could end a project's prospects of future income flows altogether, or at least substantially reduce them as better, more modern, equipment or facilities are likely to become available to replace any redundant

technology. For example, an architect could design part of a building and have it fitted out so as to accommodate a particular type of machine, or computer system, only to find out later that the systems have rapidly become technologically outmoded. Indeed whole buildings can become technologically redundant soon after completion if they are built to a very specific design and specification in order to meet a particular, now surpassed, need.

- The technique is useful in times of uncertainty. If future returns are unsure, perhaps due to economic or political instability, the quicker the initial cost of the investment can be recouped the better. By definition the longer one attempts to predict cash flows into the future the more uncertain returns become. This is especially the case with investment in property, as buildings tend to have a long economic and physical life. For example, the method could be used at the micro level when examining the payback potential of investing in double-glazing as a means of increasing the insulation properties of a building. Here, the expected payback period would be calculated based upon the present costs of installation as opposed to the savings made against current energy prices. Although estimates of future inflation in fuel costs could be made and incorporated into the calculation it is unlikely that future uncertain events would be able to be predicted. In this example there could be an unanticipated energy shortage, or conversely the discovery of a new commercially viable energy source. Both scenarios would alter the cost of energy and therefore the payback period of the investment.

The average rate of return

The average rate of return (ARR) is a method of investment appraisal designed to indicate how much of an initial investment is recouped each year. The resultant figure is represented as an average and the project with the highest average rate of return is deemed to be the most preferable. The formula used to calculate the ARR is given directly below:

$$\text{ARR} = \frac{\text{Gross Project Returns/Initial Cost of the Investment}}{\text{Life of the Project}} \times 100$$

The answer is multiplied by 100 so as to express the figure in percentage form. Again using the data from the hypothetical case study in Table 5.1 the average rate of return for both project A (ARR_A) and project B (ARR_B) can be calculated:

$$\text{ARR}_A = \frac{3250/2500}{4} \times 100 = 32.5\%$$

$$\text{ARR}_B = \frac{4000/2500}{4} \times 100 = 40.0\%$$

These results demonstrate that, on average, investing in project A will give an annual return to the investor equal to 32.5 per cent of the initial investment, whereas project B will return 40.0 per cent of the initial investment on an annual basis. As project B has a higher ARR than that of project A this method would recommend investment into project B. This preference can be expressed as:

$$ARR_B > ARR_A$$

The advantage of this technique, unlike the payback period method, is that it recognizes all cash flows over the whole life of the project. However, its general result could be misleading as it can hide important details concerning the pattern of the investment flow. For example, a global figure disguises returns that could either be highly variable or concentrated over a specific time period. Please note that the percentage figures used in this example are unrealistically high. Again they are exaggerated so as to emphasize the points raised by the explanatory example.

A conclusion on conventional techniques of investment appraisal

Both of the techniques touched upon so far have the great advantage of being quick and easy to use and simple to understand. Due to these attributes they are frequently used especially in relation to property investment at the micro level. However, a fundamental criticism of these approaches is that they do not specifically take into account the fact that, for the investor, the value of money changes over time. The time value of money is an important concept that illustrates that money received in the present is worth more than the same amount received in the future. This is simply because any monies received in the present can be invested to earn additional sums in the form of interest payments. The rise in value of a deposited sum of money in an interest bearing account can be calculated using the compound interest formula:

$$A = P(1 + r/100)^n$$

Where: A = the sum arising if 'P' is invested in an interest bearing account
P = the value of the initial deposit
r = the rate of interest applied to the account
n = the number of years that the account is held open for

For example, if a deposit of $10,000 were to be held in an account for twenty-five years at an interest rate of 10 per cent, the investment would grow to a value of $108,347.06 as seen by the following worked example:

$$A = \$10,000(1 + 10/100)^{25} = \$108,347.06$$

Such a deposit may have been made as part of a preparation for retirement. However, it is important to note that in most economies the purchasing power of the final payment is likely to be seriously eroded by inflation.

Thus, if money is not received until some future date it will have an opportunity cost attached to it. This opportunity cost is the interest payments foregone by not having immediate access to the money and therefore not being able to invest it. The longer the investor has to wait for money the greater the opportunity cost of distant returns and therefore a reduced value is placed upon monies received in the future. Thus, a technique needs to be devised that enables one to appreciate that monies received now, or at early points in an investment, are actually worth more than the same nominal amount returned by the investment in the future. This can be achieved by converting all future income flows into their present value (PV).

The concept of present value is especially important when examining investment in building projects due to the long time horizons involved in both construction and the end product itself. The conversion of future values to present values is accomplished by using discounting techniques of investment appraisal. It must again be stressed that the time value of money represents its ability to earn interest and is not a reflection of how inflation can reduce the purchasing power of money over time. The potential impact of inflation would have been assessed at an earlier stage of the investment appraisal when future cash flows were being forecasted. In this way the investor that drew up the likely returns from each project would have converted anticipated nominal returns into real figures.

Discounting techniques of investment appraisal

Discounting techniques of investment appraisal are normally considered to be superior to conventional methods due to their implicit recognition of the time value of money. This feature coupled with the long time-scale involved with investment in buildings makes discounting techniques attractive to the investor especially at the macro level. Essentially where a project involves large sums of money there is a need for enhanced accuracy. Although discounting techniques are slightly more complicated than conventional techniques they are still very easy to use and understand. The two main discounting methods are the net present value (NPV) technique and the internal rate of return (IRR). Both of these are now discussed although the former is dealt with in more detail, as the latter is merely a derivative of the first.

The net present value technique

This method of investment appraisal recognizes that the income from a project is received over a number of years and that money in the future is worth less than money held in the present due to the opportunity cost of waiting for the money. Essentially the technique converts these future cash flows by discounting them into today's present value. Thus, the net present value is the sum of all future income streams converted into present value terms minus the initial costs of the investment. In terms of a formula this can be expressed as:

$$NPV = \sum \frac{X}{(1 + r/100)^n} - K$$

Where: X = net annual income streams
K = the initial capital cost of the investment
n = the anticipated life of the project
r = an interest rate that reflects the opportunity cost of money

For example, if a project returned a net income flow for just three years before its termination, the net present value of the project could be calculated as follows:

$$NPV = \sum \frac{X_1}{(1 + r/100)^1} + \frac{X_2}{(1 + r/100)^2} + \frac{X_3}{(1 + r/100)^3} - K$$

The superscript number after each bracket denotes the applicable year for each part of the calculation. The formula generates a value that tells the investor what the project is worth in present day terms. The decision rule using this technique is that any project is worthwhile as long as the net present value is greater than zero:

$$NPV > 0$$

Moreover, as investment funds are limited projects that yield the highest NPV should be undertaken and are often ranked in this way.

Returning to our numerical example and assuming an opportunity cost of capital of 10 per cent the net present value of project A would be calculated as follows:

$$NPV_A = \sum \frac{1500}{(1.1)^1} + \frac{1250}{(1.1)^2} + \frac{250}{(1.1)^3} + \frac{250}{(1.1)^4} - 2500$$

$$NPV_A = [1363.64 + 1033.06 + 187.83 + 170.75] - 2500 = \underline{255.58}$$

Therefore, according to the first decision rule this project is viable as its net present value is greater than zero. That is to say that in terms of present value a profit is likely to be made from the investment. However, it must be appreciated that the return is not guaranteed, as all future cash flows are forecasted estimates. Note also that although in nominal terms the same amount ($250) is likely to be received in both years three and four of the project, the formula has automatically placed a lower value on the money to be received at the later date. As detailed above, this lesser present value given to the figure in the fourth year reflects the opportunity cost of having to wait a further year for returns and thus losing the ability immediately to invest them. Effectively this formula is the opposite of the compound interest formula. For example, if $170.75 were invested over four years at an interest rate of 10 per cent it would accumulate to a value of £250 (these values have been taken from the fourth year of the investment example above).

Calculating the net present value of project B produces the following result:

$$NPV_B = 669.86$$

As this value is greater than the one obtained for project A, project B would be the preferred choice in the light of limited funds. This preference can be expressed in the following way:

$$NPV_B > NPV_A$$

In order to increase the level of caution a risk premium can be incorporated. For example, a risk premium of say 2 per cent can be added to the value of 'r' that was initially chosen. This will have the effect of devaluing future income flows yet further and will thus highlight marginal projects. The more pessimistic, or risk averse, the investor the higher the risk premium in comparison with the one chosen by the optimist. Repeating the above calculations with a value of 'r' equal to 12 per cent rather than 10 will reduce the perceived net present values of both projects thereby making them appear as less attractive investments.

A problem with this method of discounting is that it is automatically biased against projects that yield higher returns in the future, as was the case with the simple conventional payback procedure. Another common criticism of this method of investment appraisal is that the investor has to select a value of 'r' that accurately reflects the opportunity cost of capital. Where this cannot be done an alternative discounting technique known as the internal rate of return (IRR) can be used.

The internal rate of return

The internal rate of return (IRR) is merely an adaptation of the net present value technique. The internal rate of return is simply the value of 'r' that sets the NPV of a project to zero. Therefore, if the construction and letting of project A were to seem viable at a value of 'r' equal to 10 per cent this may not be the case if a higher value, such as 15 per cent, had been used. By calculating a range of different values for the NPV of a project an IRR schedule can be plotted as shown in Figure 5.1. Such a diagram indicates the value of 'r' (the IRR) that sets the NPV calculation to zero for each project under consideration. The relevance of this figure, say 16.5 per cent, is that as long as the opportunity cost of capital to the investor is less than the IRR the project would be profitable and therefore worthwhile. Conversely if the opportunity cost of capital were more than the IRR losses would be made. Therefore, this procedure gives a clear numerical and visual dividing line

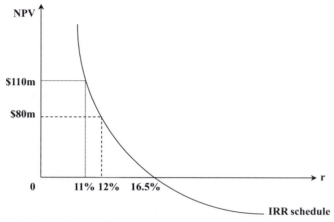

Figure 5.1 *The internal rate of return*

between project success and project failure. The line shown in Figure 5.1 may not be as exact as implied by the diagram simply because future values are, by definition, uncertain.

A conclusion on discounting techniques of investment appraisal

Discounting techniques are felt to be more accurate in that they accommodate the principle of the time value of money. However, due to their relative complexity they are not always used at the micro level, nor are they necessarily appropriate when examining very short run investment decisions.

A critique of investment appraisal techniques

It can be seen that different techniques of investment appraisal can recommend different projects to the investor. Using the hypothetical 'case studies' of the two projects in Table 5.1, one finds that the payback period method favours investment in project A, whereas both the average rate of return and the net present value methods suggest that project B promises the best investment potential. This apparent conflict of results does not mean that any one of these methods are incorrect, it simply highlights the fact that they have been designed to meet the requirements of different investors. For example, some investors may need to take a short-term view of an investment, whereas others may have the luxury of taking a longer-term view. Another issue that must be fully understood when interpreting the results of such calculations is that they are only as good as the data that is fed into them. The old adage: 'garbage in, garbage out', certainly holds here just as it does with any use of figures. Problems with the accuracy of the data used can arise for a variety of reasons as detailed below.

- Future cash flows are forecasted estimates. Even the best thought out property forecasts can be rendered inaccurate by the advent of unforeseen circumstances. For example, an unexpected military conflict in one of the main oil-producing areas of the world could push up international energy prices. Such an escalation in the price of energy can not only lead to an increase in the operating costs of a building in terms of heat and light, but it can also induce an economic recession that depresses the demand, and therefore the rental growth prospects of commercial buildings.
- The techniques assume that the life of the project is certain and known. Again, the longer one tries to predict a terminal date into the future the less certain it becomes. For example, a technological revolution could make some buildings become prematurely functionally obsolete although it may have originally been thought that they had long economic lives ahead of them.
- Discounting techniques assume a constant value of 'r' throughout the calculation. However, in reality the opportunity cost of capital to the firm is likely to fluctuate quite significantly over time depending upon alternative investment possibilities that may subsequently come to light.
- One may initially assume that one certain figure in any development appraisal would be the initial building costs. However, as buildings take time to construct, this figure can diverge from any original estimates by a considerable margin. The larger the project is and the longer it takes to construct increases the possibility of such a divergence. For example, project costs may go up because of an increase in interest rates that raises the cost of development finance. Alternatively, an unforeseen shortage in some building materials or components may increase overall building costs.

Despite these shortcomings the techniques provide useful guides and the most successful ones will be generated by the most accurate property market forecasts. These techniques are now built upon to examine how they can be used to determine the economic life of a building or a group of buildings.

The determination of the economic life of a building

If built well, the physical life of a building can obviously be very long. Indeed evidence shows that many buildings remain structurally intact over considerable time spans. However, in terms of pure financial economics it may be found that such buildings have exceeded their useful economic life after only a relatively short period. In other words, they no longer provide the owner of the asset with the maximum possible return. For example, if a technological breakthrough allowed for a new production process to be used in manufacturing, existing older factories may become obsolete, i.e. their economic lives would draw to an end. In fact, even a change in nearby land use can result in either the lengthening of the economic life of buildings or the hastening of their demise. The net result of such activity would depend upon whether or not the new land use was of a complementary or conflicting nature.

Therefore, it is apparent that investors in property need a logical framework to enable them to make well-informed judgements about the economic life of buildings in their asset portfolio. Such a framework would enable entrepreneurs to assess the potential of any changes to their property. However, throughout this analysis it must be recognized that any decisions to change a building, ranging from refurbishment to its complete redevelopment, will be constrained by a number of factors. Common constraints are in the form of planning regulations, historical preservation orders and public pressure groups. Moreover, consideration of alternative solutions to the problem ought to be encouraged before any decision is made. In relation to this last point, it should be realized at the outset that if a building shows signs of reaching the end of its economic life it need not necessarily be demolished and replaced with a new one in order to make a profit. Alternatively, extensive refurbishment or a change in use may extend the economic life of the existing building.

A variety of similar investment models are available to determine the economic life of buildings. Effectively they all propose the recommendation that:

A property should undergo some form of change when the present value of expected future net returns from the existing building, or use of land, becomes less than or equal to, the net present value of the next best use that the site could be used for, after taking all costs into consideration.

The changes that actually occur could range from the extreme of redevelopment to retaining the existing building and simply allowing it to undergo extensive refurbishment. The decision concerning what changes would be most appropriate would depend upon a variety of factors but would be chiefly dependent upon finance cost and availability as well as market demand. For example, the owners of a building may not have sufficient funds to redevelop the site, or market research may reveal that some occupiers view the occupation of a converted traditional building rather than a modern one as a viable proposition. Attractive, older buildings located in prime areas can provide a firm with an image of solidity, tradition and longevity.

In assessing such considerations the concept of present value is used, as the life of buildings is normally long and as such there is again a need to recognize the time value of money. Real figures should also be forecast otherwise the potentially erosive power of inflation could distort the accuracy and usefulness of the model over time. It is also important to note that this model is not a static one but has to be dynamic in nature. That is, it would be naïve to work out a result today in the hope that the answers or predictions would be fully accurate in twenty-five years' time, for example.

Variables within the model do, and will, change over time. Therefore, the model needs to be frequently updated in order to see if its recommendations have altered because of changed circumstances facing the building in question. Such a model could be readily run with the aid of a computer programme so as to facilitate the task, as one would merely input information regarding the new values of any changing variables.

For example, if the model were to be initially set in motion for a particular building, using both known and forecasted values for the required variables, it may suggest that, in its present form and use, the building is likely to reach the end of its economic life in thirty years. However, in the future circumstances facing the building can obviously change. For example the structure or fabric of the building may suffer from previously unanticipated repair bills. Alternatively, nearby infrastructure may be improved at a later date making the building, or the location of its site, more accessible and therefore more attractive to both users and investors. In the event of such changes the model would have to be recalculated, with redefined parameters and new data, to see whether or not the terminal date had been altered.

In order to ascertain the present value of the expected future net returns to be gained from the existing building, or use of land, values for the building's anticipated net annual returns (NAR) initially need to be calculated. Net annual returns are simply all the monies received from the building, such as rental income (gross annual returns [GAR]) minus the operating costs of that building such as the costs of management and ongoing repair and maintenance. Remember that the figures obtained for both gross annual returns and operating costs must be forecasted estimates as they occur in the future.

This relationship can be written in notation form as:

NAR = GAR − OC

Where: NAR = net annual returns
        GAR = gross annual returns
        OC = operating costs

This expression can be modified to take into account the need for discounting future values so that it reads:

$$NAR = \sum \frac{GAR - OC}{(1 + r)^n}$$

As already stated, gross annual returns are the rental received from the leasing of a building and such rental is largely determined by the level of demand for the building in relation to other relevant and available supply that tenants may choose from. As a consequence a shift in either the demand or supply curve will lead to a change in the gross annual returns subject to the timing of the next rent review. In addition to changes in the market over time there are inbuilt factors that will influence the level of gross annual returns as seen below.

- When a building is new it is likely to be perceived as being unique and state of the art. Therefore, it is likely to attract a high demand and subsequently command a high rent. However, the success of such a building is likely to encourage similar development as other entrepreneurs see that good profits are to be gained by the letting of such buildings. As a consequence of this, new and similar buildings are built which, when completed, will compete for occupants with the existing one. This resultant competition increases the total supply of floor space for the type of building in question. Thus, if demand does not rise as fast as the increase in supply, rents will eventually be driven down. Note that there may be some 'stickiness' in the fluidity of any market change as alterations to rents may only be possible during periodic rent reviews rather than continuous ones.
- As the building ages, higher repair bills could decrease rent levels over time as tenants refuse to pay high rents to occupy a building that is constantly undergoing repairs, or is in need of attention, such as elevators continually being out of order.
- Tenants may be less willing to pay a high rent for a building that is ageing for a variety of reasons other than those discussed in the point directly above. For example, the building may become technically obsolete, or the area in which it is located may no longer be suitable for its effective operation. That is, the land and other buildings in the vicinity may become run down, suppliers may have moved on or the local transport infrastructure may have changed. Such a scenario will lead to a leftwards shift of the demand curve for this type of building in its particular location.
- There is a greater degree of risk and uncertainty as one attempts to forecast further into the future. As a result valuations of potential gross annual returns tend to become increasingly more cautious the further into the life of the building that one tries to predict.
- The process of discounting future values into present value terms creates an automatic tendency for future values to decline over time.

As gross annual returns tend to decline in real terms when a building gets older, its operating costs are likely to escalate. Operating costs can increase over time for a variety of reasons as demonstrated in the following points.

- The structure and fabric of a building deteriorates physically over time and is thus subject to increasing levels of repair and maintenance.
- Older buildings are often difficult and costly to adapt to new technical requirements and demands. For example, alterations to internal configuration may be difficult to accommodate a change in use. Alternatively the building may not be conducive to the installation of modern IT equipment.
- Legislative changes concerning matters such as health and safety, fire prevention or energy wastage, could all require monies being spent on existing buildings in order to bring them in line with current and acceptable standards. For example, the removal from older buildings of

asbestos, or other materials that are now recognized as being dangerous, may involve a substantial cost. Likewise extensive measures may need to be undertaken to ensure that a building can readily be vacated in the event of a fire if fire regulations become more stringent than they were at the time of the building's design and construction.

As net annual returns are simply the difference between gross annual returns and operating costs, the net annual returns of most buildings will decline as the building ages due to the fact that gross annual returns show a tendency to fall and operating costs a tendency to rise.

However, in order to determine the economic life of a building one should not wait until the net annual returns of the existing building fall to zero. If the investor were to behave in this manner the built environment would contain a large number of poor quality buildings and urban land would not be put to its most profitable and optimum use. So as not to miss out on lucrative development opportunities, one also needs periodically to assess the present value of the next best use that the site can be put to after taking redevelopment costs into account. The estimate of the value of the land in its next best use can be expressed in equation form as shown immediately below:

$$\text{Value of next best use} = \sum \frac{\text{GAR}^* - \text{OC}^*}{(1 + r)^n} - (D + C + B)$$

Where: GAR* = gross annual returns from the next best use
        OC* = operating costs of next best use
        D   = demolition costs of existing structure
        C   = site clearance costs
        B   = rebuilding costs

When the original building was constructed the expected value of the next best use must have been less than the anticipated value of the building that was actually built. Otherwise, the alternative building would have been developed instead of the one that currently stands. However, the value of the next best use that the site can be put to is likely to rise, as one is able to build a building that reflects the changing level of the economy and technology. The demands for such a modern building are likely to be high, as was the case for the original building when it first appeared on the market at the beginning of its physical and economic life. Therefore, the stage will be reached whereby:

> *The value of the expected future net returns from the existing building, or use of land, equals, or becomes less than, the anticipated net value of the site's next best use after taking all costs into account.*

Once the current building or land use approaches this juncture one could either:

- Demolish, or change, the existing building, or land use, and replace it with another more profitable one; or
- Prolong the economic life of the existing building.

Extending the economic life of a building could be achieved by a variety of measures. First, the building could be extensively refurbished. Successful market-orientated refurbishment that was sensitive to user needs would attract higher demand for the improved facilities and working environment. Higher demand would be represented by a rightwards shift in the demand curve putting upwards pressure on rents. Second, the use of the building could be changed in order to find tenants who were willing to use the building at an attractive rent. Old industrial buildings could be converted to residential use for example. Third, operating costs could be reduced. If gross annual returns were falling such cuts could help maintain overall net annual returns. Short-term savings in repairs and maintenance could achieve this although a deterioration in the building would impair long-run net annual returns. A longer-term view could be taken by investing in improved energy efficiency in the building so as to reduce energy costs in the future.

In conclusion, this market-orientated approach is a useful decision-making aid for the property owner, manager or investor when determining the probable economic life of a building. However, as in the case of all models it is unlikely to be totally accurate on all occasions. Any model is only as good as the quality of information fed into it in the first instance. It is also important to accept that the magnitude of key variables used in the model are likely to change over time. This latter point is of particular importance when looking at investment in buildings as their economic life may span several decades if not more. Thus, it is again stressed that the model must be viewed as being dynamic and not static. Obviously the model is also continuous as any new building built, or any changes made to an existing building, will have a limited life in both economic and physical terms. In other words, the model should indicate a progression of development on a site as buildings are replaced by others as time advances.

The analysis now moves on to consider some examples of how the model can be used in a dynamic way. Specifically, a broad range of issues are examined as examples of how factors external to the model (exogenous variables) can alter the economic life of a building. Indeed the economic life of a whole group of buildings may change in that the evolution of any urban area is a continual one. Such evolution will invariably have implications for all buildings whether they are classified by type, use or location.

The economic life of industrial buildings

Due to the process of urbanization, and the importance of industrialization in that process, many inner city areas are partially characterized by the existence of old, industrial buildings. In many economies these buildings were often the result of the Industrial Revolution, but their specific location was also derived from the original, smaller urban area, and a different transport infrastructure to the one that exists today. For example, many nineteenth century factories were built around rail junctions that are now no longer in existence. Therefore, it is easy to realize that developments over time have contributed to the shortening of the economic life of such buildings although they are still often physically sound. The evidence to demonstrate this point is easy to find as one can observe old manufacturing buildings that are left derelict, are converted into offices, or are knocked down to make way for alternative land uses such as car parks.

One of the main forces at work here was the process of the decentralization of urban areas. Decentralization created the situation whereby many manufacturing firms preferred to locate on the periphery of a town rather than near its centre. For example, in many countries rail freight no longer goes to the heart of the urban area and as such industry relies upon supply and distribution by road. As a consequence, firms need to locate near the major roadways and they do not wish to be hampered by the high levels of traffic congestion so often associated with the inner urban area. Moreover, modern production techniques favour large, single-storey industrial units rather than the multi-storey configuration of pre-twentieth century buildings. The advent of the fork-lift truck and conveyor belt assembly lines were leading factors forcing this design change. By definition these modern units require large areas of land, and siting near the centre of a town would be cost prohibitive due to higher land prices there.

Therefore, the demand for these older, centrally located buildings as prime manufacturing property fell. As a result, the gross annual returns received from such buildings declined which in turn reduced their net annual returns. Moreover, the operating costs of these buildings can suddenly rise, as unanticipated repairs become more frequent. Alternatively, changes in legislation, such as health and safety legislation, can enforce expenditure upon the building such as with the provision of enhanced fire precautions. This will again lower net annual returns as an increasing proportion of gross annual returns are eroded by cost increases. In other words, the economic life of old, centrally located buildings can be shortened by developments such as the growth of an urban area, technological developments, and changes in infrastructure.

The fate of industrial buildings is not always clear. In the case of a buoyant urban economy they may be demolished to make way for a more profitable use such as retailing or office property. Conversely, if they are located in a general area of economic decline they may simply be left derelict or be demolished and cleared so that the site can temporarily be used as a car park. In this latter case it is presumably hoped by the owner of the site that a future economic recovery would bring to fruition proposals for a new and profitable development. Alternatively, especially if the building is of some architectural merit and is located in an acceptable area, it may be converted to an alternative use within its existing structure. Therefore, it is quite common to see old industrial buildings converted to residential or commercial use. In some cases though it is simply the façade of the original building that is kept. The rest of the structure is then demolished to make way for a modern purpose-built building to be constructed behind an historic frontage. The future fate of recently built industrial buildings is perhaps more obvious. Most modern industrial units rely upon a steel frame and much prefabrication. These buildings are quick and easy to pull down and are unlikely to be considered as great examples of architectural merit. Therefore, at the end of their economic lives they will be more readily replaced.

Inner city housing

As urban areas develop it is often found that that a large stock of housing remains within the inner city. This feature is despite the fact that for many years there has been a tendency for housing development to decentralize towards the more spacious and affluent suburbs. The remaining inner city housing is often of poor quality and is in competition with other land uses. The continued existence of these houses can again be explained by reference to the economic life of buildings in the ways described below.

- Houses are often kept for 'consumption' reasons rather than purely for the purposes of financial investment. As there is a non-economic motive it has to be accepted that the model is largely inapplicable. For example, elderly people may not wish to move away from a community in which they have lived for all of their lives, despite the fact that there may be superior alternative accommodation elsewhere.
- Urban planners may wish to promote, or keep, a mixed land use in the central area, with housing as an important contributory factor. As such, local authorities may attempt to enhance the economic life of centrally located residential buildings by providing financial assistance in the form of improvement or renovation grants. The existence of housing in the inner city may be continued by government policies aimed at revitalizing inner cities. For example, to reduce traffic congestion from the outlying areas government could encourage developers to develop new housing on previously built upon 'brown field' sites in the urban area rather than green field sites around the suburbs.
- In pure economic terms, the economic life of inner city housing has been extended by the private sector in a number of ways. First, landlords can increase gross annual returns by subdividing large properties into smaller units and thereby letting the building to a larger number of people. Although the rental per person is often reduced in such circumstances, the total receipts from all occupants are likely to be greater than would be the case if the building were to be leased to a single, more affluent tenant. In many developed nations there are numerous examples of three or four storey nineteenth-century houses being subdivided into several self-contained apartments. Second, landlords have maintained net annual returns by keeping operating costs to the barest minimum. The evidence of such behaviour can be seen by the poor, dilapidated appearance of many rental properties.

As has been shown, there are a variety of circumstances that can lead to increases in net annual returns that can prolong the economic life of a residential building.

Changes in urban land use

The use of adjacent or nearby land to a building, or indeed a group of buildings, will also have implications for the economic lives of those existing buildings. For example, if a complementary land use were to be developed alongside, or near, the existing buildings, the demand, and hence the net annual returns of the original buildings, could increase. Imagine older factories being made more appealing to industry by the development of improved local transport infrastructure. Alternatively, higher demand for the buildings may be stimulated by the development of other industrial units that could encourage the onset of agglomeration economies.

Conversely, the development of conflicting land uses could have the opposite effect and actually reduce the net annual returns, and thus the economic life, of existing buildings. For example, if industry were to be sited near present housing, the positive effects would be increased employment. However, on the negative side the demand for the housing may fall, as few people would wish to live in such properties if there were to be high levels of pollution and congestion created by such industry. In this case, the economic life of adjacent residential property could be reduced.

Developments in construction technology

Improvements in construction technology could hasten the end of the economic life of a building, as the redevelopment of the site becomes an increasingly attractive and profitable proposition. Changes such as improved technology in the arena of prefabrication, for example, could greatly enhance the general level of construction productivity. This would enable the landowner to replace any existing building both at lower cost and more speedily than would otherwise be the case. This would encourage the redevelopment of the site at an earlier date.

Government activity

The activities of the public sector can alter the economic life of a building, or a group of buildings, in a number of ways, examples of which appear below.

- Although a building may have technically reached the end of its economic life, planning restrictions may prevent the redevelopment of a site. Such instances are common in the case of buildings, which are protected because of their specific historic or architectural interest. In this way government intervention, through the planning process, acts as a constraint upon the model. The building may still be changed but this may be promoted through the alternative avenues of a change in use or via extensive refurbishment.
- Compulsory purchase orders (CPOs) can be used by local authorities to speed up redevelopment in some instances or extend the economic life of

buildings in others. An example of the former could simply be a compulsory purchase order being served on a building in order to make way for a new road. With such an eventuality both the economic life and the physical life of the building are prematurely ended. In the latter case compulsory purchase orders could be used to retrospectively zone out competing land uses and have them replaced by complementary ones. Such action would enhance the economic lives of existing buildings.

- Improvement and renovation grants can enhance the economic life of a building by enhancing its attractiveness to potential occupants. A rise in the demand for the building would enable higher rents to be charged which would lead to increased net annual returns.
- Interest rate policy can also have an important bearing on the economic life of a building as the cost of development finance is directly related to the ruling rate of interest. For example, if the government were to lower the interest rate, as part of an expansionary monetary policy (see Chapter 12), this could lower the costs of redevelopment and thereby may encourage early redevelopment. However, it has to be said that lower interest rates would lower the cost of borrowed funds in general. Therefore, other costly alternatives such as extensive refurbishment will also need to be examined.

The conservation of historical buildings

Increasingly societies are becoming aware of the need to conserve examples of architecture from previous ages, especially those that have a particular historical or national interest. Using the economic life of a building model in its raw profit maximizing form is obviously inappropriate here. Specifically the recommendations of the model do not take into account the external benefits that can be derived from historical buildings such as the pleasure people can gain from visiting or simply seeing such buildings. Therefore, as the market tends to ignore social costs and benefits (see Chapter 7) there is a rationale for the State to intervene in order to ensure that such buildings remain in existence. The conservation of buildings can be achieved in a number of ways as seen below.

- The building can be brought under public ownership, perhaps by being converted into a museum.
- The building could be left under private ownership and provided with a subsidy designed to support the net annual returns received by its owners.
- Buildings can be listed and once buildings are listed they cannot be tampered with without strict approval, let alone be demolished.
- One could allow, within acceptable and agreed limits, a conversion to a more profitable usage. For example, old houses could be turned into office accommodation without greatly affecting the external appearance of such buildings. In fact, as already mentioned, some developments may gain permission to merely maintain the original frontage of a building, yet completely rebuild the structure behind it.

It can be seen that the market-led decisions of the private sector, with profitability as the primary motive, go a long way to explaining the development of the built environment over time. However, it must be recognized that the public sector also plays a key role not only as a developer of public buildings and infrastructure, but also in its capacity of constraining the excesses of any private sector investment decision.

In conclusion, there are a variety of land use models that predict and explain the likely layout of an urban area, and help to demonstrate how land use patterns may change over time when subject to a variety of economic, political and demographic forces. These models primarily stem from original agricultural land use models simply because the same techniques and logic are used to describe patterns of agricultural land use and patterns of urban land use. In all of these models the concept of the bid rent curve is used as the main determinant of actual land use.

A bid rent curve represents how much an investor in, or user of, a particular land use, such as retailing for example, will be willing to pay in order to locate in a specific area. As these curves describe the relationship between the price of the land and the quality of it demanded they are effectively the demand curves for the various different forms of land use in the urban area. An important determinant of such demand is the general accessibility of the site. Examples of these demand, or bid, curves can be seen in Figure 5.2. In order to simplify the analysis only five categories of urban land use have been included in this analysis. Namely, prestige offices, retailing, middle to low income residential areas, industrial use and higher income residential areas.

Assuming that there is a need for prestige offices to occupy a central location in a town or city, perhaps due to agglomeration economies or public image, such establishments are likely to be willing to pay a high

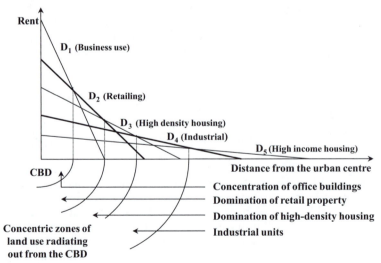

Figure 5.2   *Bid rent curves and the determination of general urban land use*

price to obtain such accommodation. On the other hand they would be less interested in securing floor space further away from this area. It is for this reason that their demand curve would fall off rapidly as one moved away from the city centre as shown by the demand curve for such properties D_1. As this demand curve reflects that these businesses are willing to pay more than any other form of land use to occupy this area, a central business district (CBD) will be created. The boundary of such office use will end, however, when an alternative use is willing to pay more for a location than these businesses are. Note that this occurs at point A as it is at this point where retailers, who also desire a relatively central location in order to attract the working populace, begin to out-bid business users as the demand curve for retailing, D_2, is now higher than that for business use. In the same way retailers are displaced when the demand curve D_3, representing the demand for high-density residential accommodation, becomes higher than D_2. One can continue this logic to determine the location of all the other forms of land use as seen in Figure 5.2.

From this, a generalized picture can be produced of urban land use radiating out from the centre. This is certainly not to say that these boundaries will be as clear as those suggested in the figure. In reality there will inevitably be some overlap between different land uses. For example, some offices may not be able to afford to locate in the central business district, others may feel that there is no need to do so, and some may not be able to find suitable accommodation in that area. It is for these reasons that offices will be found, subject to planning constraints, in other parts of the urban area. Likewise, residential land use, such as luxury penthouse apartments on the top of high-rise buildings, may also be found in the central business district. Moreover these implied strict boundaries of land use are bound to be affected by a variety of other issues, some of which are set out below.

- Geographical features may distort the actual pattern predicted by the model, the most obvious examples being a body of water such as a large river, a lake or the sea, or an area of high ground that prevents easy construction.
- Manmade physical features, for example the provision of major transport routes such as roads and railways, tend to distort land use patterns by elongating their location. The reasoning behind this is that land uses become more accessible in terms of travel time and the cost of travel if they are served with good communications.
- Legislation, such as land use planning controls, is also an important force in determining the shape of the built environment and the location of buildings. For example, the creation of an attractive green belt, or even green wedge, is one of the reasons that high-income housing suburbs can be observed on the periphery of urban areas.
- Historical reasons may produce wedges of land use that break up the general concentric pattern predicted above. For example agglomeration economics, nearness to raw materials, or accessibility to a pool of suitable labour, may have created a noticeable industrial area of a

town. Similarly, cultural reasons may produce an area of notably high-income housing that does not conform to the normal prediction of its location. This provides us with a sectoral view of land use in the urban area.

Furthermore, it must be appreciated that these boundaries tend to change over time due to factors such as the growth of the urban area, changes in transport technology and developments in information technology (on this latter point please refer to Chapter 3). These forces tend to create a number of zones of transition that are undergoing change from one form of land to another. To illustrate this point it is quite common to see in some cities areas dominated by old, large houses that were previously the preserve of high-income groups. However, low-income groups via multiple tenancies and subdivision now occupy these buildings, as those on higher incomes have moved out to the suburbs of the urban area as it has grown.

Similarly, industrial buildings that used to be on the periphery of the town and close to old transport routes may find themselves trapped well within the city as it extends over time. Due to problems of accessibility and the high cost of central urban land such industries find it difficult to survive. Many towns illustrate the existence of such properties in the form of either derelict central industrial buildings, or ones that have been converted to multiple occupancy. In fact, some imaginative local authorities and private developers have converted such buildings into retail centres or flats. This policy of change in use, refurbishment and conservation helps society retain examples of important architecture from the past. Moreover, in terms of conservation it also enables changing land use patterns to be accommodated with the minimum of redevelopment and its associated disruption to the local community.

6 Government intervention in land and construction markets

This chapter introduces various forms of government intervention that can be imposed upon the free operation of the market giving an insight into how specific construction-related markets are, or can be, manipulated by the public sector. As the emphasis is upon individual markets the following analysis can be classified under the heading of micro-economics whereas the role and impact of government macro-economic policy upon property markets is examined in Chapters 10 to 12.

Essentially, if the market were to operate without intervention it should, in the absence of any market failure, arrive at an efficient, free market, equilibrium. However, there are numerous cases where such a solution may be viewed as inequitable, or even unsafe. Indeed the market may fail altogether. Under such circumstances corrective government action, at either the local or national level, can be taken in an attempt to improve the end result. For purposes of illustration four key areas of government intervention are covered in this chapter and the next:

- Taxation and subsidization in the built environment.
- Regulatory policy: health and safety in buildings, and land use control.
- Price control policies in the built environment.
- Rectifying market failure in the urban environment.

Taxation and subsidization

Taxation

There are a wide range of taxes that can be imposed by both central and local government on both consumers and producers. The purposes of such taxation can range from the raising of revenue, to the regulation of production or activity for the good of society in general.

Taxation upon the producer

The government may impose additional taxes, or increase taxes, upon a good or service in order to attempt to reduce the amount supplied on to the

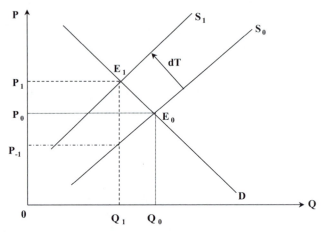

Figure 6.1 *Taxation imposed upon the producer*

market. Such taxation directed at the producer will, *ceteris paribus*, effectively increase input costs. As supply is partially a function of input costs this is likely to lead to a leftwards shift in the supply curve and therefore achieve a decrease in supply at any given price. In other words the contraction in supply will have the effect of increasing the price of the product and thereby reducing demand as seen in Figure 6.1. Here, if the market were allowed to operate freely, it would reach an equilibrium at E_0, with a price equal to P_0 and the quantity traded equal to Q_0 as shown by the interaction of the demand curve D and the supply curve S_0. However, a newly imposed, or increased, tax raises the production costs of the supplier by the amount of the tax, and therefore shifts the supply curve to the left. Thus, with a new equilibrium at E_1 less will be produced (Q_1) at a higher price (P_1). Note that normally, before the imposition of the tax, the supplier would have been willing to supply Q_1 for a price of P_{-1}, but now, to cover the tax, it charges P_1 for this level of output. Thus, the producer pays for $P_{-1}P_0$ of the tax, and the rest, P_0P_1, is passed onto the consumer via a higher selling price. The ability of the producer to pass the tax onto the consumer is largely dependent upon the elasticity of demand.

Such taxation could be seen as the driving motive behind the following examples:

- Taxing potentially hazardous or non-environmentally friendly building materials. This should have the effect of increasing their price and thus encouraging builders to use alternative, safer, and more acceptable materials.
- Increasing taxation on energy for lighting and heating such as gas and electricity. This should increase the price of such services to the householder and commercial user and thus encourage them to be more economical with the world's limited resources, as well as being more environmentally aware. Not only will higher energy bills make users

become more cautious about energy wastage, but also the tax could encourage users to install energy-saving devices such as double-glazing and improved cavity wall insulation, for example. Effectively rising energy costs would shorten the payback period of such investments and thus enhance their attractiveness as seen in Chapter 5.

● Taxing an industrialist whose factory is polluting the atmosphere. By imposing a tax on the company's product, the supply curve of the product would shift to the left, and thus less would be produced and demanded. Consumers could perhaps buy from a cheaper supplier who is using a cleaner manufacturing process, and is therefore not attracting tax. In fact, such a tax, if high enough, could encourage existing producers to invest in less polluting methods of production so as to reduce the level of tax imposed. This sort of tax could therefore lead to a cleaner built environment.

Alternatively taxes could be reduced for those operating environmentally friendly production techniques or those producing a good or service that is beneficial to society.

Taxation upon the consumer

Altering the level of tax that a consumer has to pay will obviously affect the level of income at the consumer's disposal. The higher the tax, the less the disposable income of the consumer. Thus, as disposable income is a key element of demand, increases in tax will decrease a consumer's ability to demand goods and services, and thus their demand curve will shift down to the left. Such a shift can be seen in Figure 6.2 where demand has fallen from D_0 to D_1 effectively reducing output from Q_0 to Q_1. Therefore, using the previous example, one could attempt to encourage householders to use less energy by reducing their spending power rather than increasing the price of

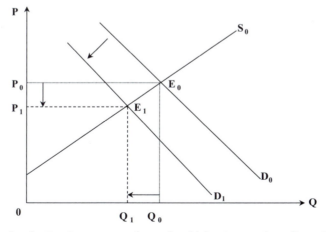

Figure 6.2 *A reduction in consumer demand as higher taxes reduce disposable income*

the product. However, the effectiveness of such a policy would largely depend upon the relative elasticity of demand between different goods and services. Moreover, the situation is complicated yet further as the consumer could continue to use the same amount of energy as before, yet cut down on other essential expenditure such as that on food and clothing. Furthermore, the impact of such a policy could be delayed by consumers taking out loan finance in order to keep their consumption levels the same as before while they find it difficult to make economies, or cannot afford to install energy-saving devices. Alternatively, lowering the overall tax burden, as seen in the debate on fiscal policy in Chapter 12, could encourage consumption.

Subsidization

By approaching problems from a different perspective government could grant subsidies to the producer or consumer in order to encourage the production and consumption of particular goods and services.

Subsidizing the producer

Monies aimed at increasing supply could be aimed at the producer who is providing a good or service that is beneficial to society. Examples could be:

- Manufacturers producing energy-saving devices or materials could be given money to encourage higher output, lower prices and further research.
- Manufacturers using methods of production which are now unacceptable in terms of pollution or safety could receive subsidies to pay for more modern equipment to replace the old, or to learn about new techniques.
- Subsidies could be used to encourage firms to provide more or more varied services, e.g. in the field of public transport, non-profit making routes that are nonetheless essential for outlying communities could be run with the help of such finance.

The impact of such subsidization can be seen in Figure 6.3. Here, a subsidy is seen to lower the supply costs of production, thus enabling producers to supply more at any given price. Under these conditions the supply curve will shift to the right from S_0 to S_1, where more will be produced, and at a lower price, as output increases from Q_0 to Q_1, and price falls from P_1 to P_0.

Subsidizing the consumer

In some cases the subsidy may be best aimed directly at the consumer. For example imagine a local authority who, wishing to improve a depressed

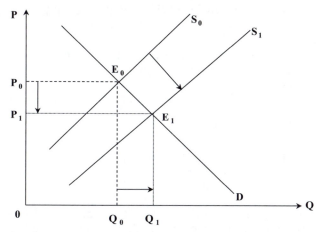

Figure 6.3 *The effects of a producer subsidy*

area, promoted an enhancement in the quality of existing privately-owned housing stock in that area. To encourage people to improve their homes, home improvement grants or renovation grants could be offered. Thus, people could obtain public money to increase the standard of their properties by undertaking improvements making it a nicer area for all to live in. Note that, as can be seen in Figure 6.4, the detrimental side-effect of this policy could be that items required for home improvements and renovation may go up in price reflecting the resultant increase in demand. The full effect of this price movement would depend upon the elasticity of supply.

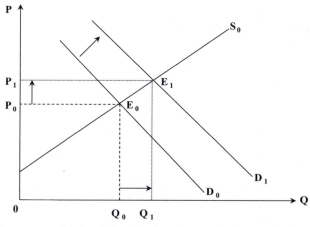

Figure 6.4 *Consumer subsidies giving rise to an increase in demand for home improvement and renovation*

Regulatory policy: health and safety in buildings and land use control

Legislation regarding health and safety in buildings and land use control are two more important areas that demonstrate where government directly intervenes in the market.

Health and safety in buildings

Whether one is involved in manufacturing, construction, the provision of property or financial services, research and development (R&D) leading to technical innovation, is encouraged so as to:

- Improve the product in order to make it more desirable for the consumer or end-user. Such improvements should help to maintain, or even increase, demand for the product, giving rise to potential positive revenue implications that should outweigh any increased costs of production, at least in the long run. For example, house builders may invest in techniques that improve the quality of their houses so as to attract more people to buy them.
- Find a means of reducing the firm's costs so that it can become more competitive. For example, the firm could find new, cheaper materials that could perform the same function as the original materials used. An instance of this would be a builder using a new, cheaper type of cavity wall insulation that had adequate properties so as to meet current energy standards.
- Enhance productivity by finding faster, more efficient, methods of production or provision of service. This should enable the firm to meet the needs of the market more quickly and therefore recover its costs at a faster rate. Such a goal is especially important where large amounts of loan finance need to be repaid by the producer so as to avoid ongoing interest repayments, as is typically the case with most construction projects. Moreover, speed of delivery is likely to attract more demand from those impressed by such efficiency.

Apart from the case of merely improving quality, all of the above actions would have the effect of shifting the supply curve to the right, as technology is a key variable in the supply function. Thus, technological change should enable more to be produced, provided, or built, at any given price. This can be seen in Figure 6.5 where the supply curve has moved from S_1 to S_2 leading to an increase in the quantity traded from Q_1 to Q_2, and a reduction in price from P_1 to P_2. As can be seen from the diagram, such technological change could be passed on, in part, to the consumer, or end-user, via lower prices. Indeed, the consumer may possibly be offered an improved product as well, although this is not necessarily the case if the technology has been designed to simply enhance profitability.

However, regulatory bodies, for example, may see the application of such technology to the productive or building process as being unsafe for the

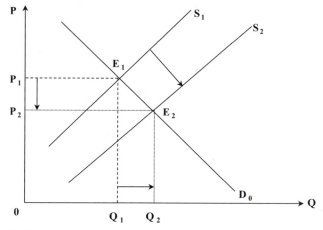

Figure 6.5 *Shifting supply due to a technological advance*

workers involved, or for the consumer or end-user. Thus, government safety regulations, building standards, or manufacturing standards could be imposed to:

● modify the new technology to make it safer and thus more acceptable
● reverse the technological change completely so as to legally prevent it from being used.

In each of these instances the supply curve is likely to initially shift back to the left, *ceteris paribus*. This retreat of the supply curve would obviously be complete in the case of the new technology being found to be totally unacceptable. It is important to recognize that many methods of production and construction can be used for a considerable amount of time before they are declared to be unsafe. Problems can be brought to the notice of regulatory bodies by the reporting or observation of frequent accidents on site, structural failure in buildings, or advances in medical technology. The detrimental effects of the use of certain materials can then be measured. Therefore, such legislative reversals, as described above, may not be immediate and could thus lead to high levels of rectification work on existing buildings.

Land use controls: land use planning

Land use planners may deliberately prevent the building, or at least limit the supply of certain types of building, so as to avoid the risk of commercial collapse, and at the same time to strive to achieve the highest possible level of welfare for society in any given area. For example, although a manufacturer may wish to locate in a particular area due to perceived advantages of location near a market, raw material or major route-ways, such development may be prevented, or restricted, if this were to produce a situation of conflicting land uses. For instance, allowing heavy industrial development in a primarily residential area, or an area of outstanding natural beauty would impose social costs upon society thus detracting from their welfare. Such social costs could include increased pollution, increased congestion due to heavy vehicle traffic to and from the factory, and the loss of green areas. Planners may also reject such an application because they are aware of other similar future developments in the vicinity. Giving all the proposals the go-ahead may lead to an oversupply of such businesses and buildings, which could subsequently lead to business failure and dereliction. Not only may jobs be lost, but also such resultant dereliction is likely to be viewed as unsightly and thus a cost to society, as well as a waste of resources.

On the other hand, planners can positively encourage the supply of certain types of buildings so as to stimulate the growth of complementary land uses. For example, planners could try to create a business or industrial park. With such developments it is argued that firms can benefit from being near to their suppliers for cost reasons, but also their proximity to competitors can enhance their awareness of the market. Such improved knowledge should facilitate discussions between firms that could lead to improved business prospects for all those concerned. These positive spin-off effects are known as agglomeration economies. Note that the land use control can occur at both the national and local level. On a national basis certain guidelines can be laid down so as to achieve a cohesive national socio-economic policy, whereas local governments are concerned with examining individual cases as well as adhering to an overall central policy. For a further examination of the economic rationale for land use planning refer to the debate on market failure in chapter 7.

Price control policies

For a variety of reasons central, or local, government may feel that the pure market result for some goods and services is an unjust one. As such, they may be able to seek the power to impose and enforce either:

● a maximum price control; or
● a minimum price control.

Both of these policies are individually examined below.

Maximum price controls

A maximum price control is the setting of a legal maximum price on a good or service whereby the providers of those goods and services are not allowed to charge above this imposed price. That is, a price ceiling is created and prices are not permitted to go through, or higher than, this artificial constraint. Such a ceiling would be imposed if the government believed that the market price of a particular good or service was inequitably high when looked at from the angle of the general consumer. For example, in the case of a good, the market rent for rented accommodation may be deemed to be too high for those on low incomes. In the same way, it may be felt that current charges for some services in the house buying process are too high to be fair to the home buyer. Thus, the government could place limits on the fee level for a valuation or structural survey needed to obtain the necessary finance for house purchase. By definition the price ceiling would have to be set at a price below the market equilibrium in order to be effective.

However, it is important to note that, as with most policies, price ceilings rarely achieve their goal without causing secondary effects, which may be positive or negative. To demonstrate this, two applied examples are now examined in some depth. Specifically the examples chosen highlight additional measures that need to be imposed in conjunction with such price interference in order to reduce or eliminate any potential ill-effects of the original policy.

Example 1: The rented residential accommodation market

If the rented housing market were allowed to operate freely, a series of equilibrium rents would be achieved for the various types of rented residential property available. Therefore, for sensible analysis the market for such property should be subdivided into smaller sub-markets to reflect the variation in the types of accommodation and their specific location. For example, it is likely that a small apartment for rent in a town centre will attract a different clientele than a large rented house in the suburbs.

For the purpose of illustration, the case of small rented apartments is now examined. In urban areas characterized by such properties it would be

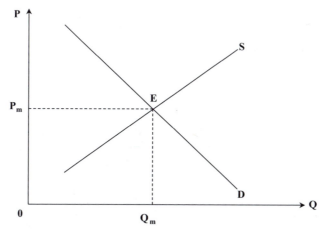

Figure 6.6 *The free market for apartments*

expected that there will be a demand for such properties and a potential supply of them as shown by the demand curve D, and the supply curve S in Figure 6.6. The demand curve, as usual, will be downward sloping since fewer will wish to pay rent for such accommodation if it is priced too high, yet such properties would attract many if the rent is sufficiently low. For example, a low rent could attract young people to move away from their parents' home as they could afford the independence of their own house to rent. Likewise, low rents are likely to attract people on relatively low incomes, such as students, who could afford to live with fewer people in each dwelling.

On the supply side, the supply curve would be expected to exhibit the normal upward sloping relationship between price and quantity. That is, there will be some landlords who are willing to rent their houses at a low rent, whereas others will only consider it if rental income is to be relatively high. Thus, in Figure 6.6 a market equilibrium rent of Pm is achieved and Qm apartments are rented out. As with other markets, this equilibrium will not necessarily remain stable as either, or both, of the curves are likely to shift over time. For example, demand could increase for such properties if there were an increase in student numbers at a local college or university. Or, on the supply side, supply could be reduced if a change in tax legislation made rental income less attractive to landlords. These landlords are likely to react to the resultant reduction in the profitability of renting by withdrawing their properties from the market.

Returning to the original market equilibrium rent of Pm, the local government, or city council, may feel that this is an unjustly high rent for the less well-off communities who may wish to live in such accommodation. In these circumstances the concerned authorities could impose a maximum price control, or price ceiling, on such dwellings. For example, it could decree that landlords were not allowed to charge more than a rent of Pc as shown in Figure 6.7. This price ceiling would legally prevent the price

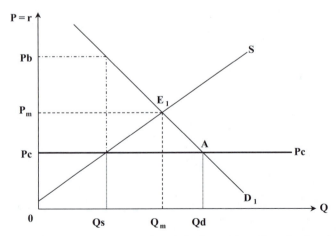

Figure 6.7   *Rental property (apartments) and the imposition of a price (rent) ceiling*

(rents) rising above Pc to the market price of Pm. Obviously such a maximum price control is only sensible and effective if the maximum price is set below the ruling equilibrium price. If the ceiling were to be set above the equilibrium rent, rents would simply adjust back to the market clearing level, as they would be free to move as long as they did not exceed the price ceiling. That is, once the policy is in force, prices are only free to move at or below the artificially imposed ceiling.

The policy-makers may initially feel that they have solved the problem of high rents for the local community as rents have been formally lowered. However, by looking at Figure 6.7 one can predict that a situation will arise whereby the demand for such accommodation increases from Qm to Qd as rents have decreased from Pm to Pc. In other words, renting an apartment is increasingly feasible for many due to its reduced cost. On the supply side lower rental income will produce a tendency for less accommodation to be supplied, as fewer landlords are willing to let their properties for such a low rent. Thus, eventually, supply could contract to Qs from Qm, giving a net excess demand equal to QsQd. Such a decrease in supply, however, may not be immediate, as existing tenants will normally have some form of short-term security of tenure. That is, landlords could not evict tenants as soon as the policy was imposed, as contractual obligations normally decree that they would have to provide reasonable and adequate warning of the termination of any lease agreement.

Therefore, it would appear that a maximum price control policy by itself could produce quite substantial negative effects as detailed below:

- Although the local authority has succeeded in decreasing the rental price for the dwellings from Pm to Pc, this is likely to result in a situation where less accommodation is available than was previously the case. Note that the amount of accommodation available on the market has dropped from Qm to Qs.
- To exacerbate the shortfall above, the level of demand has increased from Qm to Qd.
- Low rental incomes could mean that landlords spend less on their properties in terms of repair and maintenance. Moreover, they may shelve any plans for improvement. Such actions would obviously lead to a deterioration of the quality of such housing stock.
- The imposition of artificial price constraints could unintentionally promote an illegal market. Under such conditions potential tenants may be tempted unofficially to offer the landlord a price above the price ceiling rent in order to obtain one of the limited number of apartments available. If such trading were to arise, only the relatively well-off could afford to rent an apartment as the illegal market price (rent) would be as high as Pb if there were only Qs properties available to rent. Alternatively, another undesirable outcome could be over-crowding as many people start to live in each apartment so as to share the burden of the high rent.

Therefore, as can be seen from the analysis above, policy-makers need to attempt to rectify the situation of excess demand by introducing additional policies in conjunction with the initial price control. Examples of such policies are suggested below.

An attempt could be made to shift the supply curve to the right. Ideally this could be done in such a way as to ensure that the new supply curve intersects with the original demand curve D_1 at point A in Figures 6.7 and 6.8, where a new equilibrium could be achieved. To facilitate this increase in supply the public sector could construct and provide more dwellings for rent. However, apart from the time that it would take to construct such buildings, it would require a substantial initial capital outlay that may be beyond the housing provision budgets of many local authorities. In addition, in the long run, ongoing costs would be incurred for the management, and repair and maintenance of such properties. Figure 6.8 shows the new supply curve S_2 passing through point A.

Government may attempt to encourage more to be supplied given the existing supply conditions as represented by the original supply curve S_1 in Figures 6.7 and 6.8. That is, they could introduce additional measures to encourage the suppliers of rented accommodation (landlords) to supply more dwellings at this lower rent of Pc. Specifically Qd properties need to be supplied. This could be achieved by offering the landlords a subsidy equal to PcPs as shown in Figure 6.8. However, this new policy measure imposes significant ongoing costs of subsidization as represented by the area PcPsBA. Arguably, however, this subsidy could be reduced to a differential subsidy shown by the area ABC in Figure 6.8. The theory behind this view is the observation that according to the original supply curve S_1 some landlords are willing to accept a lower rental than others. For example, the upwards slope of the supply curve implies that the landlord represented by Qs is quite happy with a rent of Pc, whereas it is only the landlord represented by Qd who insists upon a rent of Ps before supplying their property on to the market. Therefore, in theory, different landlords could be given different subsidies reflecting their own individual situation. In reality, however, this would probably be impossible to administer, as few landlords

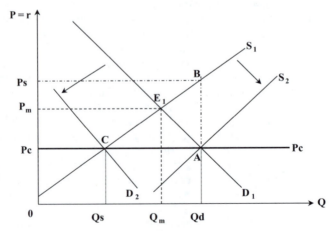

Figure 6.8 *Rent control with additional policies*

would admit that they were happy with a rent below Ps as they would all wish to receive the maximum subsidy available.

Alternatively, one could attempt to solve the situation by attempting to reduce demand in this market. Ideally demand should be shifted down to the left from D_1 to D_2 so that the new level of demand intersects the original supply curve S_1 at point C. Here a new equilibrium could be established as can be seen in Figure 6.8. Such a shift in demand could be achieved by encouraging people from this rented market to move to another housing market. For example, the government could tempt people to buy their own home. In this way former tenants would enter the owner-occupied market by moving out of the rented sector. To stimulate such a shift in demand central government could lower interest rates, or increase (or introduce) mortgage tax relief, for example.

If the initial aim was to reduce the rent for existing tenants from Pm to Pc, one could make the tenants initially pay the full rent of Pm, but give them a subsidy in the form of a rent rebate equal to PmPc. A problem with this policy is that in keeping the cost of rental accommodation down it is likely to encourage a further increase in future demand, and subsequently therefore, an increase in the number of applicants wishing to receive a rent rebate. This obviously represents a substantial ongoing cost for those administering the policy. Moreover, as demand rises this may lead to a widening gap between the official rent ceiling Pc and the potential market rent. Such an outcome is due to the upward sloping nature of the supply curve and the potential shifting of the demand curve to the right. It is partially for this reason that the level of the price ceiling needs to be reviewed periodically to ensure that it is achieving its objectives in a sensible manner.

Finally, after imposing the price ceiling policy-makers could additionally enforce strict long run security of tenure for existing tenants. Therefore, despite the receipt of a low rental, landlords could not evict their tenants. However, as mentioned earlier, this is likely to lead to a running down of the physical condition of such properties as landlords cut costs in an attempt to re-achieve their original profit margin.

Example 2: Price control on building materials

Imagine a situation whereby a developing nation's government is concerned that in order to facilitate rapidly rising economic growth, more building development needs to occur so as to ensure that sufficient housing, offices, factories, and warehouses are available for the smooth running of the economy. In order to achieve this goal, the government may feel that it needs to encourage existing building firms to build more, and for new construction firms to join the market, so as to increase the supply of new buildings. One way that the government could attempt to achieve such a goal of increased output would be to lower building costs and thus increase the potential profitability of all firms in the industry, remembering that the main incentive to supply is the attraction of making profit.

In order to reduce costs, the government could lower taxes levied upon building firms. However, as this would adversely affect revenue to the government they may alternatively attempt to reduce the costs facing building firms by placing a price ceiling on essential building materials, such as cement. Again it can be seen that, if introduced in isolation, such a policy will lead to potentially detrimental outcomes which need to be reduced, or rectified, via additional policies introduced in conjunction with the initial price ceiling. (It should be noted that governments, which adopt a centrally planned system of economic administration by regulating prices in the economy, may inadvertently place a price ceiling on goods if fixed prices are not reviewed and adjusted regularly enough.)

By using Figure 6.7 again, but in the context of the market for cement, it can be seen that the immediate impact of the price ceiling policy is to lower the price of cement from Pm to Pc. Such a reduction in price will tend to increase the demand for the good from Qm to Qd as desired. However, the suppliers of cement will now only wish to produce Qs, as more production would not be profitable given the new conditions of a regulated price. As such, marginal cement production would be shut down leaving an excess demand of QsQd. If this situation were allowed to persist an illegal market would be likely to arise whereby builders have to offer suppliers higher than normal prices (up to Pb) to obtain some of the limited supply. Moreover, such a situation is often exacerbated as some hoard cement in the hope that further artificial shortages will drive prices up even further so that they can make even larger profits on the resale of the good. Thus, less building work is likely to occur, as an essential raw material becomes very costly and largely unattainable. In an attempt to rectify this situation the government could introduce several additional policies to support the initial price control. The following discussion concerns such additional policies and should be viewed in conjunction with Figure 6.8.

Government could encourage an increase in supply so that the original supply of cement as represented by the supply curve S_1 shifts to the right to S_2. This could be attempted in a variety of ways:

- Allow imports of the material. The problem with this policy however, is that it uses valuable foreign exchange, and could encourage a dependency on imports.
- Attract more firms to set up in the production of cement. However, this is likely to be difficult, as few firms would find joining the industry an attractive proposition when the price of the good is so low. Therefore, such a policy may have to be promoted by offering firms subsidies or tax incentives, both of which would detract from government revenues.
- Nationalize the industry and enforce greater output (only a strongly interventionist government would introduce this). Unfortunately, however, past evidence has shown that many publicly-controlled industries throughout the world have had a reputation of poor organization leading to higher production costs and even lower output in terms of both quality and quantity.

- Finance research, or provide research facilities, to find cost-cutting methods of production, or other productivity improvements for the existing firm(s) in the industry. This would have the effect of enhancing the producers' profits thus encouraging higher output.
- Encourage the existing supplier(s) to supply more by offering them a subsidy equal to PcPs. Alternatively, it may be possible to administer a differential subsidy that could be distributed to the firm if it had different plants with different circumstances each requiring a different incentive to make them profitable. The area ABC represents such a differential subsidy.

Alternatively, the demand for the good could be reduced preferably so that the demand curve shifted to the left from D_1 to D_2. This could be achieved by trying to reduce the reliance upon cement by changing the method and type of construction. For example, an increased use of alternative methods of construction such as more prefabrication is a possibility.

The government could directly subsidize the building firms for every bag of cement used. Initially this would require a subsidy equal to PcPm, although this figure could well increase as demand for the product increased due to its effective price decrease.

Finally, if supply had fallen to Qs, the government could introduce a rationing system so as to distribute equitably the limited supply. However, such systems are often expensive to introduce, and hard to effectively police. Moreover, in this instance, it defeats the original objective of encouraging more supply and more building work.

Minimum price controls

A minimum price control, or price floor, is a situation where governments decree that people cannot charge, or offer, below a certain official minimum price. Therefore, such policies are initiated when the government believes that the price determined by the market is too low. In such a case, public intervention is thought to be desirable so as to enforce a higher price greater than the market equilibrium. For example, we can see this in many countries in the form of minimum wage legislation for certain types of labour. That is, governments enforce a legal minimum wage that must be paid by employers. To have the desired effect of increasing wages such a minimum wage, or wage floor as it is sometimes known, must be set at a level greater than the market wage.

This situation is shown in Figure 6.9 where Pf represents the wage floor, a wage set above the market level of Pm. If the market operated freely, workers (for example manual, building site labour) would receive a wage (w) of Pm, and Qm people would be employed. However, if a wage floor, Pf, were imposed for such labour, wages would have to legally increase to this new, higher level. Such increased wages are likely to attract more people into the industry due to the higher rewards obtain-

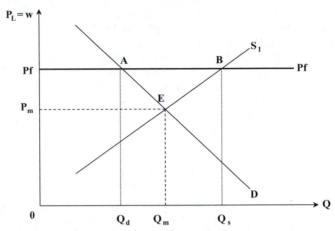

Figure 6.9   *Minimum wage legislation and construction site labour*

able. Those that join the industry could be either formerly unemployed or were working in another industry that now pays comparatively less. Given the current supply curve S_1, which represents the potential supply of construction site labour in our example, Qs, rather than Qm, workers would now offer their services at this higher wage.

However, with such high wages, employers are likely to seek ways of laying some labour off so as to reduce the rising costs of employment. A reduction in the labour force may be possible as it becomes more economic to replace some labour with more capital-intensive methods of production, or firms insist on higher productivity levels from the remaining workers. Indeed, improved site management could lead to a reduction in the numbers required. It is also possible that not only will firms contract by laying off workers, but also some weaker firms could go out of business altogether if they fail to cover these increased costs. Thus, one is left with a situation of an excess supply of labour equal to QdQs. In other words there is unemployment in the industry. Therefore, just as with price ceilings, there is a need to introduce additional policies in order to ensure the success of the initial objective. Such accompanying policies could be:

- To encourage a reduction in the supply of labour to the industry so that the supply curve S_1 shifts to the left causing it ideally to cut the demand curve at point A in Figure 6.9. This could be attempted by offering government retraining schemes to retrain the workers to work in another industry.
- To encourage existing employers to employ more workers by giving the employer an employment subsidy. The ideal result of this would be to increase the demand for labour so that the new demand curve cut the supply curve at point B in Figure 6.9. A subsidy of PmPf would be needed to achieve this.

The cost of such policies would be largely determined by the relative elasticity of the demand and supply curves. For example, the demand curve for labour could be made perfectly inelastic from Pm to Pf if workers agreed to increase their productivity in exchange for the higher wages. In such a situation nobody would be made redundant as a result of the initial policy. If this were the case, supplementary policies would not be needed.

7 Market failure in the built environment

Many argue that if the formation of the built environment were left entirely to the workings of the private sector and the market many undesirable results would occur. As a consequence the villages, towns and cities that we live in would suffer from a variety of inadequacies. This standpoint stems from the belief that the market fails in a variety of circumstances in the urban environment. The term market failure is normally used to categorize these problems. This is certainly not to say that the market is rejected completely as a useful mechanism in which to shape urban areas, it is simply stating that it must be recognized that the market may produce sub-optimal results in some instances.

Once these areas of market failure have been recognized it is hoped that careful public sector intervention via the use of urban economic policy could solve, or at least reduce, their adverse consequences. In fact, as will be seen shortly, both land use planning and building regulations, for example, can be viewed as public sector responses to market failure. However, it must also be said that public intervention is not necessarily always the correct solution to market failure as government policy itself has been known to exacerbate difficulties, or at least create undesirable 'spin-offs' from the original policy. In other words, far from curing the problem, public initiatives can either fail to solve a problem, make it worse, or create others that were not there in the first instance. Thus, the following text will take a critical and analytical standpoint when examining the role of urban economic policy as a means of correcting market failure in order to achieve the optimal result in the formation of urban areas.

Several areas of potential market failure can be identified in the built environment. These are listed below before the text goes into a detailed examination of each in turn.

- externalities
- public goods
- monopoly situations
- imperfect information
- society's decisions.

Externalities in the built environment

Externalities occur when the costs or benefits, from a particular good or service, to society as a whole, are not adequately reflected in the market price for that good or service. It is important to realize that examples of both negative externalities and positive externalities can be observed in the built environment. Negative externalities come about because of the detrimental impact on society of certain actions by others, whereas positive externalities arise in situations where society benefits from the actions of such third parties. However, most texts and articles normally concentrate on the former category of externality as it is the negative aspects of living which unfortunately normally generate the most controversy and are seen as being more emotive areas of discussion.

As externalities are not reflected in market prices they are said to exhibit non-pricing. Moreover as externalities have implications for people in society other than the direct user or users of the good or service, they are also said to be characterized by interdependency. It is for this reason that externalities are frequently referred to as spill-over costs in the case of negative externalities, and spill-over benefits in the case of positive externalities, as interdependency implies that people are affected by the actions of others.

In conclusion, externalities arise because there are a number of instances when consumers or producers will make decisions in the light of their own costs and benefits (internal costs and benefits) that will produce an impact upon the welfare or output of others in ways that are not reflected in the prices facing those consumers or producers. It is felt that some form of public intervention should be used to address the issue of externalities as it is feared that if the market were left to its own devices it would create too many instances of negative externalities, and would encourage too little activity that promoted the existence of positive externalities. So as to understand this issue more clearly, the text now goes on to briefly analyse the cases of both negative and positive externalities using examples drawn from the built environment. After this discussion, various forms of related public intervention are examined, and the analysis concentrates upon town planning as the example best suited to demonstrate the relevant issues.

The case of a negative externality

As seen above, a negative externality is created when the behaviour of an individual or group of individuals adversely affects the welfare of another individual or group of individuals. Such problems are not captured by the market mechanism as the producers of the ill-effects do not have to pay for this aspect of their actions, and those who are affected by them receive no compensation in the absence of public intervention. The list of such negative externalities in the modern built environment is sadly almost an endless one, but some of the more obvious ones are listed below:

- Air pollution created by industry located in or near the built environment. Although industrialists obviously have to pay for the productive process itself, in the absence of public intervention to reduce the emission of harmful pollutants, they will not have to pay for the damage created by the spill-over effect (pollution) of their activities. Similarly, unless there is intervention to the contrary, air pollution created by vehicles driving in the streets is not a cost borne by the motorist. The costs of air pollution on the urban society ranges from its detrimental health effects, to the need to clean buildings that are aesthetically damaged by such emissions. Indeed many forms of air pollution carry with them the additional problem of creating an obnoxious smell.
- Noise pollution can be generated from a variety of sources such as noisy neighbours that continually host rowdy late-night parties, heavy industry and motorized traffic, for example. However, it is perhaps the latter example that is the major creator of this form of negative externality in most urban areas. Again, it may only be with public intervention that some form of noise control could be placed on offenders. For example, the motorist may not be too concerned about the noise from the engine of their car when they are stuck in traffic outside an office building, as it is the occupants of the building who are suffering a loss of welfare. Likewise, in a purely profit-orientated, market world, the industrialist may not be too concerned by the noise levels of the productive process as long as that particular method of production maximized the firm's profits. Such noise created by industry is not only detrimental and harmful to the people who work in the factory, but it could also be a nuisance to people who live nearby especially if the factory operated on a twenty-four hour basis. Even if the noise from the productive process itself was not so great, traffic noise created during the change-over of shifts, as well as that of delivery vehicles arriving and leaving the factory gates could be a sufficient source of concern.
- Water pollution. Again, manufacturing industry is a good example of how water pollution can be perceived as a negative externality. Factories that discharge untreated, or partially treated, waste into waterways can poison, or at least reduce the quality, of the river and thus damage the welfare of people who may derive a benefit from using the waterways. Such affected parties may include anglers and swimmers, but there may be other industries wishing to use the water from the river, who now have to incur the additional costs of cleaning up the water so that it is of an adequate standard to be used for their productive process. Similarly, arable agriculture activity can lead to water pollution as surface run-off water drains large quantities of fertilizers into the river which can lead to the rapid growth of harmful algae killing off some forms of aquatic life.
- Visual pollution created by the construction of an unsightly structure or a building inappropriate to its surroundings.
- Traffic congestion. Private road users obviously have to pay for the running costs of their vehicles such as fuel, servicing, insurance and depreciation. However, in a pure market economy, other costs of vehicle

use that are imposed upon society would not have to be met by the motorist. For example, the more people use their vehicles the more the urban environment will suffer from the associated problems of air pollution (see above). Moreover, congested streets also impose costs on the urban population in a number of other ways. For example, congestion increases the difficulties for pedestrians who wish to cross roads; increases the number of vehicular accidents, especially those involving pedestrians; and creates vibration damage to buildings whose structures and foundations have continually to absorb the onslaught of heavy traffic. This latter point is especially important in the case of older, historic buildings whose structures may be weakened by the passing of time. Both their designers and builders could not possibly have envisaged the modern transport structure and technology of the present day and the damage that it can cause from associated factors such as acid rain and vibration.

These examples are simply a brief examination of some of the main negative externalities common to the urban environment, but they do not represent an exhaustive list. Recognizing that negative externalities exist, and are an issue for those living in urban areas, they can now be examined in more detail so that policy can be formulated to reduce or even eradicate such problems. In order to do this, the subject of economics provides a simple model that should help measure and quantify the extent of the problem. Although this model is extremely versatile and could easily be used to examine and illustrate all of the above examples, this part of the book will only deal with two examples due to the confines of space. The two examples that have been selected, one examining the case of urban traffic congestion, the other looking into the case of pollution in the urban environment, have been chosen to show the flexibility of this model in its ability to analyse two seemingly diverse forms of negative externality. It is important to note that although the model attempts to quantify the extent of such externalities, this process is very difficult and unlikely to be fully accurate in practice. For this reason the model's use is impaired, but it should not be dismissed as at least it highlights the fact that externalities exist and are thus worthy of further consideration and action.

Figure 7.1 can be used as the basis for both of the chosen case studies. In dealing with the example of traffic congestion first, the diagram can be seen to examine the case of a busy urban road. The model depicted in the diagram looks at traffic flow per hour and the costs to both the private user and to society as a whole created by every additional vehicle that uses the road. The demand for the road by motorists is shown by the demand curve D. This demand curve illustrates the benefits derived by the road users from its use. In particular the demand curve should reflect the marginal private benefit (mpb) derived from such road use. Note that the demand curve exhibits the normal inverse relationship between price and quantity demanded. In other words the higher the effective price of using the road, the less people will be encouraged to use it. However, if there were few alternatives to this route, one would expect the demand curve to be highly inelastic. As society derives no additional benefits from such road usage the

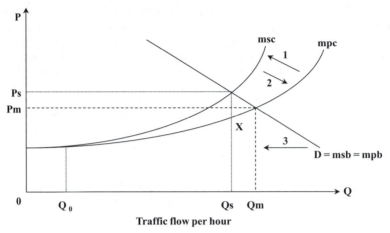

Figure 7.1   *Measuring a negative externality*

marginal social benefit (msb) curve will be no different to the marginal private benefit curve. In other words: msb = mpb = D.

The actual costs to the private road user are shown by the marginal private cost (mpc) curve. These costs not only include the financial outlay involved in running the motorist's vehicle, but also reflect the cost in terms of time taken to travel along the road. That is, marginal private costs also take on board the notion of an opportunity cost to the individual as people have to spend time in their car in traffic rather than being elsewhere, either at work or at leisure. At low levels of traffic flow per hour, for example Q_0, such private costs are relatively low, as the journey is both quick and easy. However, as more and more vehicles use the road, traffic builds up and congestion can occur. Once the rate of traffic flow has reached this stage, private costs are likely to increase for a number of reasons. First, more time is lost at work or at leisure as the journey time is increased due to slow-moving traffic. Second, fuel consumption is increased, as vehicles are rendered more inefficient by the continual stopping, or slowing down, that occurs when driving on a busy road. Moreover, the mechanical wear and tear on a vehicle is increased under such driving conditions. Thus, if left to the market, road users would use this particular road until the cost of using it outweighed their demand for it. Such a point would be reached at Qm traffic flow per hour. Beyond this point traffic congestion would be deemed so severe that road users would seek alternative routes, use the road at less congested times of day, or reduce non-essential, marginal trips.

However, although the market system has arrived at a solution, many would argue that because of the existence of externalities associated with high levels of traffic, such a solution would be an incorrect one. The argument is that the overall costs to society of the use of the road are different to those of the motorist as they also include externalities that are not captured by the market mechanism. Therefore, the costs to society of each extra vehicle using the road are additional to marginal private costs and are represented by the marginal social cost (msc) curve. Because these marginal social costs are additional to marginal private costs the marginal social cost curve is higher than the marginal private cost curve.

At low levels of traffic flow such as Q_0 the costs imposed upon society are relatively small, as traffic is free-flowing and relatively slight. However, as congestion builds up the additional costs of air pollution, increases in road traffic accidents, and the nuisance value of not being able to cross the road easily, are all examples of increasing costs upon society. Thus, the model suggests that social costs increasingly diverge from private costs and would rise to an optimal traffic flow such as Q_s. Beyond this point the number of vehicles using the road would be unacceptably high if the costs imposed on the urban public at large rather than just the private road user were also taken into consideration. In other words, the differential between marginal social costs and marginal private costs becomes too great.

Therefore, if the public sector's aim is to increase the welfare of society rather than the welfare of the individual, policies need to be implemented that could:

- Increase marginal private costs so that they come in line with marginal social costs (see arrow 1 in Figure 7.1). A policy of increasing the price of road usage to the private motorist from Pm to Ps should reduce the demand for that road as exhibited by moving back along the demand curve D. If this rise in costs was achieved the use of the road would be at the social optimum of Qs traffic flow rather than the private optimum of Qm. The resultant welfare gain is simply the saving in social costs that would have occurred if traffic flow were to go beyond Qs. A variety of policies could be attempted in order to increase the cost of road usage so that marginal private costs are increased and are more in line with marginal social costs.
- Decrease social costs so that the marginal social cost curve was effectively eradicated or brought down to coincide with the marginal private cost curve (see arrow 2 in Figure 7.1). In this way the private optimum of Qm would not be deemed as being problematic.
- Reduce the level of demand for the road so that the new demand curve passed through point X on the diagram (see arrow 3 in Figure 7.1).

Examples from each category of policy are given below.

- Charging for the use of the road at strategically positioned toll stations. Such a charge would effectively close the gap between marginal private costs and marginal social costs by increasing marginal private costs.
- Similarly, a system of electronic direct road pricing (DRP) could be introduced in order to increase the costs of using the road. This would be achieved by either the use of roadside (off car) or in car meters that would bill the driver of the vehicle for use of the road. Such charges could be graded so that they reflected busy times as opposed to other times when congestion was typically slight.
- Fuel price costs could be increased in order to raise indirectly the cost of using the road. This would again cause an upward shift of the marginal private cost curve.

- Engineering solutions can be attempted in order to encourage the demand for this congested road to decrease, preferably so that the new demand curve passed through point X as it is at this point that the desired traffic flow of Qs would be achieved. Such an engineering solution could be the construction of a ring road around a town that would remove the need for many vehicles to pass through the town and use the road in question. Alternatively traffic-calming measures, such as the introduction of speed humps and more pedestrian crossings, could discourage some from using the road. On the other hand, facilitating the flow of vehicular traffic on the road by improving the road itself, perhaps widening it, could lower marginal social costs. Such an approach would delay the onset of the point where congestion began to occur.
- Public transport could be improved to offer car drivers a real, viable alternative to using their car. Again this would produce a leftwards shift in the demand curve.
- Methods of reducing or eliminating some of the social costs could be introduced to bring the marginal social cost curve down towards the marginal private cost curve. The control of harmful car exhaust emissions can be reduced via the installation of catalytic converters on all cars. In addition petrol that contains lead can be replaced by an unleaded alternative.

Exactly the same model could also be used to analyse the case of the negative externality of pollution. Imagine that there is a factory located within, or near to, the urban area that is producing goods, and that the manufacturing process it uses releases a degree of pollution into the atmosphere. The demand for the factory's output is illustrated by the demand curve D. Again the normal inverse relationship between price and quantity demanded would be expected. That is, if the firm tried to charge too much for its product consumers would attempt to use less of it, or seek alternatives, and thus quantity demanded would fall.

The marginal private cost curve in this instance represents the costs of manufacturing the product. Such costs are likely to rise as output increases as more raw materials are used and more labour is needed. Thus, the market will tend to produce up to the point where the marginal private cost curve cuts the demand curve, and therefore Qm output is produced. However, because of the resultant creation of air pollution society may deem that such a level of output is too high as it could be damaging to the health of local residents for example.

In order to ascertain the socially optimal level of output, and therefore the optimum level of pollution, from this factory the demand curve would need to be equated with the marginal social benefit curve. This latter curve also includes the costs imposed upon the community by the negative externality of air pollution. Such a social optimum would thus be given by the lower output of Qs rather than Qm. Note that some pollution is acceptable to society as the marginal social benefit curve has already begun to diverge from the marginal private benefit curve by the time we reach Qs. These low levels of pollution are presumably endured as they pose no great danger to health, nor do they create any other significant problems.

The case of a positive externality

Positive externalities arise when the consumption of a good or service generates benefits to those other than the actual consumers themselves. For example, imagine that an individual, or group of individuals, improved the quality of their own housing. Such improvements could be of the form of ensuring a high level of repair and maintenance of the buildings themselves, or the landscaping of the grounds that surround them. Although this may be done for their own private benefit such as improving their living environment, and increasing the value of their homes, it may also have positive spill-over effects that benefit society as a whole. In other words, expenditure on items for the purpose of improving the quality of housing in an area could enhance the visual and environmental aspects of the area for those who visit, work, or pass through it, as well as improving the housing standards for the residents themselves.

Just as with negative externalities, the case of positive externalities can be examined by formulating a simple economic model that should enable one to go some way to quantifying the degree of the issue in question. Such a model is depicted in Figure 7.2. In this diagram it can be seen that in the case of a positive externality the marginal social cost curve is the same as the marginal private cost curve. Therefore, msc = mpc. The reason for this is that the production of such benefits imposes no additional cost to society itself and thus no adjustment needs to be made to the curve. The curve does rise however as improved housing standards can only be achieved at a rising cost.

The demand for such home improvements by homeowners is represented by the demand curve mpb as this reflects the marginal private benefit of such improvements to them. Therefore, owner-occupiers will invest in housing improvements until the marginal cost to them (mpc) is equal to the marginal private benefit that they receive. Thus, in this example homeowners will tend to consume a house quality of Qm at a price of Pm. However,

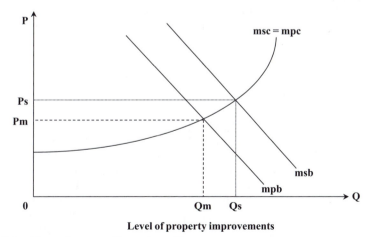

Figure 7.2 *Measuring a positive externality*

as argued above, the benefits to society of such home improvements are likely to be greater than the benefits simply accruing to the individual. This can be demonstrated by showing that society's demand for this work is higher than the demand of individuals. Here it can be seen that society's demand, as represented by the curve msb, is significantly higher than the demand curve of individuals.

The problem is that, if left to the market, housing quality would only be as high as Qm although society would gain optimum satisfaction from a housing quality of Qs. Thus, it is argued that policy should be formulated to encourage individual owner-occupiers to improve their housing beyond the point where the marginal private benefit curve cuts their marginal private cost curve. This could perhaps be achieved by the granting of home improvement subsidies or renovation grants to cover the extra costs required to bring housing up to the socially determined level. Effectively this would have the impact of raising the demand curve by the desired level. In this case society would experience a welfare gain, as these are improvements that would not have occurred without public intervention.

Urban policy and externalities

It has been recognized that if the market were left to itself to shape urban areas it could create an abundance of detrimental spill-over effects (negative externalities) that would reduce the overall welfare of people living in the urban environment. Therefore, it could be argued that carefully designed and administered policy could be aimed at reducing or eliminating such negative externalities so as to improve upon the outcome of the market. Just as importantly it must also be accepted that the market on its own may not encourage a sufficient level of positive externalities. Therefore, it is again felt that government policy could be aimed at promoting the maximum number of such positive spill-over effects so as to improve the environment of the urban population. Government intervention in this field can be at both the national and local level. So as to appreciate the types of policies that could be imposed a brief analysis of some of the main policy initiatives is examined immediately below.

Taxation and subsidization

Governments could attempt to tax the producers of negative externalities. For example, if there was a factory creating noise and air pollution due to the type of productive process used, the firm could be taxed to encourage it to switch to an alternative manufacturing process that created less negative spill-over effects. If the firm were to use a new, approved technique it would not attract the tax. Alternatively, the government could attempt to promote positive externalities by encouraging activities that could create positive spill-over effects. Such activities could be promoted by making them more affordable with the assistance of public-funded subsidies.

For example, if there was a part of the housing stock that fell below socially acceptable norms, the government could offer subsidies such as home improvement grants to encourage owners to improve their accommodation. This would be done in the hope that not only would such action improve the living standards of those who lived in the area, but would also increase the welfare of others in the community via the creation of positive externalities.

These approaches are sometimes referred to as attempts to 'internalize externalities' as the policy-maker is trying to incorporate externalities into the overall costs and benefits of society. For example, the internalization of negative externalities would try to bring marginal private costs in line with marginal social costs.

However, there are difficulties with this approach to the externality issue. Firstly, the relative desirability and undesirability of such externalities need to be assessed in order to decide which ones to take action on. People in society are not a homogeneous group, and some will have different value judgements to others. Secondly, it is often difficult to place a monetary valuation upon externalities, and therefore assessing the correct level of tax or subsidy will be a resulting difficulty. Thirdly, it needs to be ensured that the system of taxes and subsidies is both flexible and responsive to changing circumstances, and that it should also be relatively economical to run.

The setting of standards

The government could aim to impose standards in the hope of improving society's welfare over time. For example, if a polluting factory failed to meet current standards the authorities would be given the legal right to impose fines on such offenders. In a similar way, firms could be forced to undertake activities that promoted positive externalities such as landscaping the surrounding area of their premises, and ensuring the upkeep of their buildings. If firms failed to reach such standards they could again be fined. The setting of standards for the construction of new buildings (building regulations) can also be used as a means of ensuring that negative aspects of poorly designed or badly built buildings are reduced, and that the positive aspects such as easy access and good natural lighting, are maximized.

Physical controls

Physical controls could be imposed by government to ensure that acceptable criteria are being met. The authorities could insist that anti-pollution devices, such as filters, could be fitted to factorys' chimneys that were creating dangerously high levels of toxic emissions.

Land use planning

The planning process could be used in order to remove the creators of negative externalities from certain areas of the built environment, and promote land use patterns that stimulated the maximum range of positive externalities in other areas. If the other measures listed above failed to eliminate the problem of pollution caused by industry in a certain area planners could either:

● Remove the offenders, or generators, of the negative externality perhaps via the creation of an industrial park in a certain area of the town or city that was sufficiently far away from competing land uses such as residential areas.
● Remove the sufferers of the negative externality by relocating households to a cleaner, more pleasant area of the built environment that was specifically zoned for residential and complementary land uses such as retailing and leisure.

Such relocation of land uses could be both a costly and time-consuming process in an existing urban environment, yet would be relatively easy to achieve in the formation of a new town. Essentially planners would need to reduce the amount of conflicting land uses, and encourage complementary land uses wherever possible through a policy of zoning. Importantly, such factors as the direction of the prevailing wind should also be taken into consideration so as to ensure that, as much as possible, the pollution from the industrial area does not blow back on to the residential area and thus nullify the effects of the zoning policy.

Bargaining

In situations where the externality, the creators of externality, and the parties affected by the externality, can be easily defined, a solution may be made possible by the concerned parties discussing the issue together. For example, if one group is suffering from the effects of pollution from a particular factory, they may be able to negotiate some form of compensation from the factory owner. If their cause is a just one, support can be given to the case from the legal system as the law is essentially designed as a form of protection for society. Moreover, consumers can give their cause considerable weight by organizing themselves into a pressure group perhaps boycotting the firm's products if it does not agree to the demands of the people. Such consumer action has met with varying degrees of success in the past and largely depends upon the product in question. For example, a heavy armaments factory will not be affected by a consumer boycott as it is the military rather than the general public who are the clients of such output.

In conclusion on the subject of urban policy and externalities it can be seen that a variety of policy measures can be, and have been, attempted in order to tackle the externality issue. Specifically the public sector is attempting to improve upon the solution given by the market. However, as already suggested, the public sector must ensure that their policies are sufficiently flexible and well administered in order to achieve this goal. Failure to do this could at best result in the policy failing to achieve its desired objectives or, at worst, the policy could inadvertently make the situation worse.

There are also those who believe that the market will eventually take externalities into account. Such a belief negates the case for intervention in the market. It is argued that in an area that suffers from a high degree of negative externalities house prices will reflect this by being lower than would otherwise be the case. Therefore, people are compensated for these detrimental spill-over effects by being able to purchase property at a low price. Conversely, areas that benefit from a high degree of positive externalities are often characterized by high house prices. Therefore, the market mechanism can be seen as a method of highlighting the externality issue through price. Moreover, there is the suggestion that if people choose to live in an urban area they must accept the existence of a certain level of negative externalities.

A conclusion on externalities

It is important to remember that externalities can be both positive and negative in nature, and that each category is equally important. It is also useful to appreciate that externalities can be identified and categorized into four main types depending upon the source of the externality and the party that it affects. These categories are listed and briefly described immediately below.

Producer/consumer externalities

These are perhaps the most frequently discussed form of externality primarily because they are normally very noticeable. An example of such an externality is the now well-used case of the factory polluting the atmosphere. Here we have a producer (the factory) creating a negative externality that adversely effects the welfare of the inhabitants of the area around the factory because of air-borne waste. Likewise if the factory was causing the pollution of waterways it would reduce the pleasure derived by people who use such waterways for recreational purposes such as swimming or boating. Similarly it may be argued that an architect in designing and constructing a building that is not in keeping with its surroundings, may impose the negative externality of an aesthetically unpleasing eyesore which is not only unpleasant in itself, but could also detract from the whole surrounding area. Producer/consumer externalities need not only be negative in nature, however. An example of a positive externality being created by a producer to the benefit of the consumer is perhaps seen in the world of agriculture. Here the farming of the land creates a well-managed countryside that is pleasant to see and to walk in when allowed.

Producer/producer externalities

Producer/producer externalities occur when the activity of one producer adversely affects the production of another. The pollution created by heavy industry could reduce the yields from surrounding agricultural land, or pollute nearby waterways. A positive externality can perhaps also be imagined under this heading. If, for example, there were two office blocks adjacent to one another and one building was to undergo extensive external refurbishment it is likely to improve the look of the area in a way that could also attract clients even to the unimproved building.

Consumer/producer externalities

Consumer/producer externalities occur when the actions of the general public have an impact upon the welfare of producers. Ramblers and dog walkers who visit the countryside may inadvertently wander off the designated footpaths and cause damage to agricultural fields. Such activity will obviously damage the farmers' return from such fields. Similarly, the industrialist may have problems of ensuring sufficient access for lorries into the factory if nearby roads are clogged with parked vehicles of workers in other factories or offices on nearby access routes.

Consumer/consumer externalities

These occur when the behaviour of one individual, or group of individuals, affects the welfare of another individual, or group of individuals. An example of consumers creating a negative externality would be a neighbour continually hosting noisy, late-night parties. Alternatively, an example of a positive externality under this classification would be the homeowner whose garden was carefully landscaped and their house kept in an immaculate state of repair. Such actions are likely to improve the welfare of all those that see the house rather than just that of those who live in it.

The case of public goods in the built environment

Public goods are goods and services that are valued by individuals, but are unlikely to be provided by the market mechanism, or at least in sufficient quantities. Common examples of public goods in the urban context are public open space, such as urban parks and street lighting.

The reason that the market does not tend to cater for such items is that they arguably exhibit both the characteristics of non-rivalry and non-excludability. Non-rivalry implies that one person's consumption of a good does not impair the potential utility that another individual can derive from that good. For example if one person walks past a street light the amount of light available to others is not reduced in any way. Likewise it could be argued that other people using the facility of a local park does not affect one's own enjoyment of such public open space.

More significantly though, public goods also exhibit non-excludability. Non-excludability implies that once one provides the good one cannot exclude anyone from using it, and if this is the case it makes it difficult, if not impossible, to charge a price for it. For example, once street lights are turned on there is no way that their use could be limited solely to those people who have paid a contribution for the service. In other words, the problem of the free rider can occur, where a free rider is an individual who derives a benefit from a good or service that they have not paid for, while others have. It is chiefly for this reason that the only way that such services can be financed is via a legally enforceable system of taxation rather than direct charges.

In order to clarify this issue, the case of public goods can be directly contrasted with that of private goods. Private goods exhibit both rivalry and excludability inasmuch as you can exclude others from using your private property. Moreover, your enjoyment of your private property largely excludes others from using it. Furthermore, as you can determine the use of a private good, you could charge for its use if it were to be used by others. For example, your home or your own private car are both illustrations of these points.

Therefore, it would appear that in order to ensure that urban areas contain a sufficient quantity of public goods, direct government intervention in the market is required to raise the finance for their provision. Thus, it would seem that town planners must decide upon how much urban road space to provide, how many areas should be designated as open public space, how much street lighting should be provided, and so on. In order to do this planners must accurately determine the demand for such goods so that their correct allocation is provided in terms of attempting to maximize society's welfare.

However, just as there are supporters of the view that the market can address the externality problem, there are those who argue that the assumption that certain goods should be classified as public goods is not necessarily a valid one. The proponents of such a view stress that for a good to be a pure public good it would need to fully exhibit the dual characteristics of non-excludability and non-rivalry. However, finding goods with both of these characteristics is relatively difficult, and if this is the case the argument for public goods begins to weaken. For example, urban roads are normally seen as being the responsibility of local government as they are assumed to be public goods. However, if the issue is examined more carefully, the conclusion could be drawn that urban roads do not conform to either of the characteristics necessary for the classification of a public good. With respect to non-rivalry it is not true to say in this case that one person's use of the road does not affect the utility of others using the road. It is quite clearly the case that, as congestion begins to build up, using the road becomes slower and more dangerous to name just two negative aspects (see the negative externality debate above).

More importantly, to be a pure public good, the good must adhere to the characteristic of non-excludability. However, it is possible to exclude certain people from using the road by either charging a toll or only admitting vehicles with a special licence, for example. Once this is done, a price can be charged for its use, and therefore there is the opportunity for the private sector to get involved in the provision of goods and services that have traditionally been assumed to be in the public domain.

Monopoly situations in the built environment

Monopolies are most commonly associated with the production of goods and services (see Chapter 9), however the issue of a monopoly can occur in the case of urban land and development. For example, imagine an area of land that is currently made up of four individual plots (plots A, B, C and D), each with an existing building and owned by four separate owners. It may be the case that this block of land is the ideal site for the development of a new, larger building, perhaps a hotel that would replace the existing structures. If the existing buildings were in a poor state of repair, dilapidated or functionally obsolete, and had no historical interest, it may well be in the interests of the urban society that the new development were allowed to go ahead. Increased employment opportunities are an obvious benefit.

In order for the new building to be constructed property developers would need to acquire the block of land in total and would thus need to negotiate selling prices of the existing buildings and sites with the current owners. It may be relatively easy at first to buy up the plots as each existing owner is in competition with the other landowners to sell their property. So, for example, landowner A may not wish to ask for too high a price just in case they lose the sale to one of the other owners B, C or D. In fact such behaviour is likely as long as there is competition to sell amongst owners, and as long as the existing landowners are unaware of the developer's overall intentions of purchasing all four properties.

However, if the developer has managed to purchase all the plots except plot D, for example, and the owner of that plot knows that the developer needs that plot in order to make the development viable, he or she could hold out for a ransom price. In other words, the owner of plot D could ask a price well in excess of the selling price of the other three properties as this remaining landowner is now in a strong monopoly position.

In fact, according to the above logic, it is in every one of the property owners' interests to be the last seller in such circumstances as they try to hold out for the higher price, and there is therefore the danger that no one will sell at all. Thus, the market may not be able to cope with this problem and monopoly landowners could be in a position to delay or prevent a beneficial project being undertaken. Or one owner may gain significantly more from their property than those who sold previously in competition with one another.

Therefore, it is argued that there is a need for public intervention, such as the power of a compulsory purchase order (CPO), which is designed to enable a local authority forcibly to purchase properties that are required for a particular worthwhile development. Thus, such a system could be designed to facilitate the assembly of units of land and property that are presently in fragmented ownership. Without such powers many feel that much urban investment and redevelopment would not take place at all.

However, planners must ensure that such powers are not granted too hastily as a solution could be achieved via the refurbishment of the existing buildings rather than via their demolition. Perhaps one of the reasons that many cities have lost fine examples of historical architecture is that planners have too readily redeveloped areas without much consideration to the conservation of the built environment.

Another argument against the granting to the state of such interventionist powers is that they intervene too directly with the freedom and wishes of the land or property owner. A property owner may not wish to sell in the current market if prices are depressed, and thus they may be waiting to sell in more prosperous times where a higher selling price can be achieved and therefore more profits can be made. However, the serving of a compulsory purchase order would force the owner to sell at the current, undesirable, market rate.

Also, in relation to the theory of the firm, planning powers may also be used to try to ensure a socially desirable mix of buildings on a site. For example, the owner of a vacant plot of building land set aside for residential development may find that it is likely to be more profitable to restrict the development to a few highly-priced, executive-style houses rather than any other form of residential buildings. However, the needs of society in general may not be met by the building of such houses as they will only cater for people on high incomes, or for companies who wish to purchase such houses for their senior management, for example. The requirements of the local community may be better met if the development consisted of smaller houses or a mix of types of houses that included smaller units such as starter homes. Therefore, the planners should recognize the undesirability of the market solution in such instances and insist on a broader development that caters for a range of income groups.

Imperfect information in the urban market

If a market were to operate perfectly all those in the market, both buyers and sellers, would have perfect information concerning the ruling level of prices, the behaviour of competitors, and so on (see the theory of perfect competition in Chapter 9). However, in the absence of any public intervention in the built environment such perfect knowledge is extremely unlikely, and if this condition of perfect knowledge does not hold the results produced by the market mechanism are unlikely to be satisfactory. When examining perfect knowledge in the case of the producer, two potentially damaging scenarios could occur:

- Developers could underestimate the demand for a particular type of building so that the urban area becomes inadequately serviced by certain land uses. For example, if the developer did not appreciate that a town was going to experience an increase in wealth or an increase in in-bound migration, it may not provide sufficient retail areas. Therefore, at least in the short run, there will be increased pressure on existing facilities that may lead to overcrowding and congestion just to mention two detrimental effects of overusage.

- Developers may overestimate the demand for a particular type of building. If a development firm notices that its rivals are constructing and successfully selling office buildings in a town, it may also wish to take advantage of such potential profits. However, if such behaviour is left unchecked the market will become flooded with a particular type of building. Such an oversupply will tend to drive down rental levels of the existing buildings as clients have greater choice, but it could also mean that many buildings remain unoccupied and are at risk of becoming derelict. Vacant buildings tend to attract a whole host of other social problems such as vandalism and other related crimes. Alternatively, consider the example of a developer constructing a large retail centre, who is unaware that another similar facility is going to be built by another firm in the same vicinity in the near future. If both retail centres are built, consumer demand may not be sufficient in the area to ensure the long-run prosperity of two facilities. Just as with the case of the oversupply of office buildings, many retail units could fail or remain unoccupied and this would reduce the benefits derived from such centres. However, supporters of the market system could respond here by arguing that such competition will cause a price cutting war amongst producers that will be to the benefit of consumers. This argument is often backed up by the view that restrictive planning creates unnecessary scarcity that in turn causes urban prices to remain high.

Advocates of the market mechanism may also argue that the market would eventually arrive at the correct provision, as the weaker developments would fail in the long run leaving the urban area with the correct quantity of each type of land use. Buildings that were originally intended for office use, but remained unoccupied, could perhaps be converted in order to accommodate a change in use and be used for another purpose such as education. However, this argument perhaps fails to appreciate that the short-run adjustment period may be quite a long one, and that firms may not supply less profitable buildings that are still of value to urban society. Therefore, there is a case for public intervention in the form of planning, whereby planning would be used in conjunction with the operation of the market in order to achieve a socially acceptable range of developments.

It is likely that planning departments have the best knowledge of the future requirements of the urban area. This is because they are involved in the planning of residential areas and transport infrastructure, which both have a bearing on the location and success of other land uses such as commercial and retail centres. By definition planners need to forecast the future growth and changes in the urban area, and such studies and ongoing monitoring is invaluable in relation to ascertaining the correct number and location of future developments.

The supporters of the free market system may react to this view by stating that this argument for centralized planning assumes a lack of sophistication on behalf of private firms. If private development companies wish to ensure their success and maximize their profitability it is presumably likely that they too would attempt to forecast the future requirements of an urban area before committing themselves to any costly development. In fact many large development firms do indeed have their own forecasting personnel or departments. However, this standpoint still does not address the provision of less profitable land uses. Moreover, it is still unlikely that the private sector would be able to obtain as much information as the public sector with respect to the overall view, and therefore it would seem best that the two were to work in conjunction with one another.

A firm of retailers wishing to open an outlet in a town might be unaware of the plans of its competitors as other firms may also wish to set up in the vicinity. Therefore, in a pure market situation this firm could be expecting a highly profitable location, but in the light of subsequent competition the site could turn out to be highly marginal or even unprofitable. Such uncertainty could put investors off a town, which could result in a loss of potential jobs, and desirable land uses. With a planning system in force, however, the firm could look at the overall future structure of the urban area and be able to make more certain decisions. Moreover, the planners would know how many other applications they had for similar projects.

An example of imperfect information and the consumer can be seen in the case of the house buyer. With reference to town plans a potential house purchaser can quickly see the other types of land use that are in the immediate area of a particular house. Such knowledge could prevent house buyers buying a property that will soon be adjacent to potentially undesirable land uses such as heavy industry. In a pure market situation such information would not exist in an easily accessible, centralized form. Instead the location of development would depend chiefly upon the profitability decisions of individual firms and such decisions and locations could change rapidly in response to continual changes in the market giving rise to an air of uncertainty.

Social views and the urban environment

Many argue that the preferences of the individual expressed through the market mechanism are not the same as the views of society as a whole. The simple explanation of this view is that individuals may try to maximize their own welfare without too much regard for others in society. It is because of this view that planning can be seen as adopting a paternalistic policy which guides society to make decisions that will maximize the welfare of the whole urban community rather than that of the individual.

There are those who believe that if everyone maximizes their individual welfare this should aggregate to the maximization of society's welfare. However, this apparently logical view fails to recognize that the market only caters for those who can pay (the whole notion of effective demand). Therefore, there are bound to be sectors of the community, for example the poor and homeless, who will not be included in such a system, and subsequently such people could not benefit from its operation. Moreover, people in society may not fully appreciate the value that they derive from certain land uses such as open public space, and therefore if left to the individual, urban areas could suffer a lack of certain important amenities,

e.g. public parks. Similarly, planning permission for a residential development could be refused if the planners felt that individuals had failed to
recognize the full significance of the nuisance value created by nearby
industry, or a major road.

Furthermore, society may not fully take into account the needs of future
generations. The preservation of historical buildings leaves an important
record of architecture and the development of communities over time. A
purely market-orientated urban environment may demolish such buildings
if they are no longer seen as being profitable. Finally, it is argued that society
will not cater for a sufficient level of provision of merit goods. Merit goods
are those goods which it is felt that society should consume to an adequate
level in order to ensure an acceptable level of welfare for all. Examples of
such merit goods include schools, libraries and hospitals.

A conclusion on market failure and the rationale for public intervention in the urban market

It can be seen from the above analysis that there are arguments both for and
against the market as being the sole driving force behind the creation of the
built environment. Generally it would appear that there is some merit in
instigating a degree of intervention in the market in order to correct its
inherent shortcomings. Without such intervention the welfare of society as
a whole is unlikely to be maximized. However, as mentioned above, such
intervention would have to be managed carefully and flexibly in order for it
to achieve its desired results. Research has revealed that there are many
fundamental and common criticisms of the planning process, and indeed
other forms of public intervention, in the built environment. Problems with
the planning process are briefly discussed below.

- It is argued that planning is all too often used in a negative manner that
  prevents development. Although one of the roles of planning is, of course,
  to prevent the construction of buildings that are unsightly or not required,
  many feel that planners rarely take on their other role of positively
  promoting worthwhile development.
- Many complain that the planning process is a highly bureaucratic one
  that leads to the slow processing of case-by-case applications. Such time
  lags can be highly costly to the developer as it lengthens the period of
  development and the potential need for borrowed funds. Therefore,
  development costs can be increased and the receipt of their revenues
  postponed as the sale of the completed building is held back due to delays
  in obtaining planning permission at the early stage. Such difficulties
  could well lead to the cancellation of investment in the urban environment, and any cost increases from delay certainly mitigate against the
  development of marginal projects.
- Another common argument is that some of the guidelines of planning are
  too broad and lack sufficient flexibility to accommodate individual cases
  with specific circumstances. Although it is felt undesirable to have a

certain type of building in a particular area of town there may be circumstances where it could be seen as acceptable. A ban on industry near a residential area perhaps needs to ensure that it differentiates between the cases of light industry as opposed to heavy industry, for example.

- Some studies of town planning suggest that planners sometimes fail to appreciate certain knock-on effects created by their actions. For example, they may have overlooked the fact that their decision has had a distributional impact upon an area of the town. It may be the case that giving permission for the development of a new retail centre encourages higher income groups to live in that area. This may cause house prices to rise beyond the reach of poorer members of the community who wish to live in that area because their family and friends are residents there.

Therefore, just as there are criticisms of the market, there are equally strong arguments against various forms of state intervention. However, hopefully it can be seen that the problems of the planning process above could be rectified by making such intervention more efficient and streamlined. Finally it must be re-emphasized that the relative merits of all systems should be examined and hopefully the best from each should be taken. If this view is adopted we are likely to see the public and private sectors working together to achieve the maximum possible welfare for urban society.

Part 2
Applied theory of the firm

8 Applied cost and revenue theory for construction and property

Whether examining firms in the manufacturing or service sector of the economy, applied cost and revenue theory is primarily concerned with assessing how much output firms should produce each year in order to achieve their financial objectives. If examining a firm of house builders, for example, the theory that now follows will give guidance to that firm as to how many developments it should initiate and continue under varying market conditions. Alternatively, it could give a firm of surveyors advice as to how many client instructions it should accept in a given period. Importantly, users of the theory need to recognize that continual market fluctuations require an ongoing updating of calculations and analysis if the firm is to be successful. It should also be noted that the following discussion concerns the normal situation of a firm facing a downward sloping demand curve. In some instances firms may be so insignificant in terms of overall industry output in an area that they become price takers and are subject to a completely inelastic demand curve. For a description of such cases please refer to the theory of perfect competition and its application in Chapter 9.

As a logical starting point it is assumed in theory that the rational firm in the private sector wishes to maximize profits, whereby profits can simply be calculated by subtracting all of that firm's costs (total costs) away from its total receipts (total revenue). This can be written in abbreviated notation form as:

$$\pi = TR - TC$$

Where: π = profit
TR = total revenue
TC = total costs

In fact, it is hardly surprising to find that, in reality, profit maximization does seem to be the aim of most private firms. However, empirical evidence does show that some firms are more interested in sales maximization rather than profit maximization. Maximizing sales may be due to mere company ego, or because of long-run ambitions of future profitability. A firm pursuing a policy of sales revenue maximization may be attempting to initially 'flood' the market with its product or service, so as to make a name and reputation for it as a leading brand or provider. After this initial objective has been achieved the firm could then go on to follow profit maximizing procedures.

At first it might logically be felt that the more a firm produces and sells the more profits it will make. However, this assumption is only true up to a certain point as profitability can fall if a firm produces too much as can be seen from the following analysis. In addition it should be noted that provision from the public sector might be based upon other non-monetary goals such as the maximization of social welfare.

To find out how much a firm should produce in order to maximize its profits, the potential revenues that the firm can receive and the costs that it may face needs to be examined. Importantly it must be recognized that anticipated costs and revenues could change over time as the market inevitably changes. Initially many may feel that there are many new terms and concepts to learn in order to comprehend the theory introduced in this chapter. However, it should soon be appreciated that such a fear is largely unfounded, as all of the terms below are interrelated. Moreover, if one takes the time to think about the meaning of the terms themselves they are largely self-explanatory.

Revenues of the firm

Total revenue

Total revenue (abbreviated to the initials TR) is the total amount of money that a firm receives from selling its good or service. Thus, total revenue received by the firm is the obtained price of the good or service multiplied by the particular level of output that the firm manages to sell on to the market. Therefore, total revenue received by house builders can be calculated from the actual sale price of their houses on completion of contract multiplied by the number of houses that they actually sell. Alternatively, revenue received by a firm of surveyors is calculated by multiplying the number of surveys conducted by the relevant fee level. Total revenue can be expressed in notation form as:

$$TR = P \times Q$$

Where: P = price
         Q = quantity

As total revenue is concerned with a relationship between price and quantity the potential revenue facing the firm can be estimated by examining the demand curve for that firm's products or services at the present time. Or, if more appropriate, the demand curve that the firm is likely to face when its product or service comes on to the market. As already implied, it is important to appreciate that such a process should be under continual review as market demand can change over time.

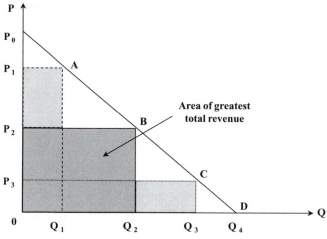

Figure 8.1   *The demand curve and total revenue*

Using the example of the house-building firm, imagine that it has estimated the likely demand for new starter homes in a town and that its findings are represented by the demand curve in Figure 8.1. As suggested above, this demand curve may need to be a projection of the level of demand in the future as it will take some time to get the product, in this case housing, on to the market. Time will be spent assessing the market in the first instance, obtaining planning permission, finding suitable building land, preparing the site, organizing production, marketing the development and so on.

The perceived demand schedule suggests that it is anticipated that the firm will sell none of its starter homes if it set the price too high, say at a price of P_0, and therefore no revenue would be received. This result is because nobody looking for a starter home would be willing to pay such a high price as normally cheaper alternatives would be available from the second-hand market, or people would be attracted to the rented sector instead. On the other hand, if the firm thought of restricting output on the site to a low number of units (Q_1), for example, it could estimate from the demand schedule that it could sell the dwellings at the relatively high price of P_1. This high price would be possible only if no near substitutes were available at a lower price that potential house purchasers may purchase instead. If there were no near substitutes available, high prices would prevail because it would then be likely that there would be fierce consumer competition for the small number of houses available on the market (see Chapter 1 for a discussion on the determination of market equilibrium). However, note from the area OP_1AQ_1 in Figure 8.1, that although a high selling price may be achieved, overall total revenue would be low as so few units would have been produced and sold.

Alternatively, by constructing more houses, say Q_2, the level of competition between potential purchasers is unlikely to be as great as there is more choice, and therefore the market will only clear at a lower equilibrium price of P_2. However, despite a lower selling price, more houses could be sold and greater total revenue received. This is shown by the fact that the area of total revenue generated at the lower price of P_2 is given by the area OP_2BQ_2, which is larger than the area of total revenue gained at the higher price of P_1 of OP_1AQ_1.

Building more houses though will not necessarily increase total revenue further. This is simply because if too many houses come on to the market, the supply curve for such housing will shift to the right and prices will be driven down. In other words, there would be an increased level of competition between the sellers of such houses, both old and new, to attract the limited number of people interested in starter homes as represented by the demand curve. Therefore, if more houses were built, say Q_3, the price would be driven down to P_3. Such a substantial drop in price would lead to low revenues despite the large number of houses sold. Note that the area OP_3CQ_3 representing total revenue received from the sale of these houses is less than area OP_2BQ_2, which represents the monies, received from the sale of fewer, Q_2, homes. In fact if too many houses are built there may be nobody interested in buying them if there is not sufficient demand in the area, and as a consequence some buildings would remain unoccupied. Figure 8.1 suggests that in the case of our limited example, at the theoretical extreme, even if the houses were free only Q_4 would be occupied as there would be a finite number of people looking for such a house in that area. Moreover, a parallel consideration is that if a site is too densely built on it may become less attractive to potential purchasers who could decide to look elsewhere for a better planned, more spacious development as a place to live.

Therefore, if the firm is faced with a downward sloping demand curve, as is likely to be the case, one will find that total revenue initially rises with increased sales, yet it will reach a peak and decline again. This conclusion and its supporting information is transferred from Figure 8.1

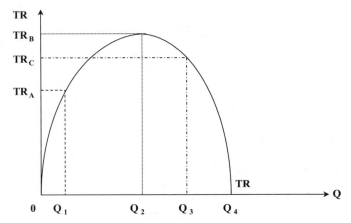

Figure 8.2   *The total revenue curve*

to Figure 8.2 in that the areas under the demand curve in Figure 8.1 give the data for total revenue that enable the plotting of the total revenue curve in Figure 8.2.

Before leaving this introductory discussion on total revenue it is important to note that total revenue is not the same as profit; it is simply the total monies received by the firm. Therefore, at this stage, the firm still does not know how many houses to build or how much profit it will make. To discover comparative profitability at each level of potential output the firm also needs to obtain precise information concerning its costs. Moreover, a slightly more detailed analysis of revenues can be useful to the firm as well.

Marginal revenue

The concept of marginal revenue (abbreviated to the initials MR) is a logical extension of the principles discussed above. Marginal revenue is simply obtained by measuring the change in total revenue caused by selling an additional unit of output. This can be expressed in notation form as:

$$MR = \Delta TR$$

Where: Δ = a change in

If nothing is sold, no monies would be received and thus total revenue would be zero. Then, as the firm begins to sell some houses, total revenue obviously increases, and the change in total revenue becomes positive. However, with the given demand conditions, the addition to total

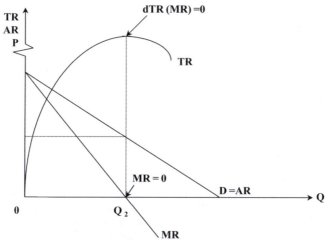

Figure 8.3 *Marginal revenue derived from total revenue*

revenue being created by more houses being sold will become less and less as prices have to be lowered to sell the greater level of output. In fact, one would reach the stage where the price of the good (houses) had to be dropped so much in order to sell all available output, that additions to total revenue (marginal revenue) become negative. Once this point has been reached the total revenue is at its peak and then begins to decline, as can be seen in Figures 8.1, 8.2 and 8.3. These diagrams show that if the firm is faced with a downwards sloping demand curve, the more it sells the lower will be the selling price of its output, and therefore its marginal revenue curve will also be downward sloping. For example, it was seen that in order to sell Q_3 houses instead of Q_2 the price would have dropped from P_2 to P_3 to ensure that all the houses are sold in the prevailing market conditions. In other words, marginal revenue declines as output increases.

Average revenue

Average revenue (AR) is simply how much the firm receives on average. Thus, mathematically, average revenue is total revenue divided by the current level of output sold and can be expressed as:

$$AR = \frac{TR}{Q} = \frac{P \times Q}{Q} = P$$

From this it can be seen that average revenue is the same price as at any given level of output. It therefore follows that the average revenue curve is the demand curve, as it is the demand curve that indicates at what price the goods (houses in this instance) will be sold. Thus:

$$D = AR$$

To understand the distinction between average revenue and marginal revenue note that the additions to total revenue (marginal revenue) are zero at Q_2, yet average revenue is still positive at P_2. This is because although houses are still being sold, and revenue is still being received, the price is so low that total revenue would have been more if the firm had tried to sell fewer houses at a higher price.

Conclusions on revenues

The above analysis is straightforward, but it must be realized that its accuracy hinges upon the ability to forecast the exact location of the demand curve as it is from this information that all revenues are derived. Moreover, demand can be quite volatile in many markets and thus subject to continual change. As an example, housing demand can easily be affected by frequent fluctuations in the interest rate.

Costs facing the firm

For a firm to find its optimal level of output, with respect to making the maximum possible profit, it must also discover what costs it faces. In order to do this the firm needs to work out the cost of producing every level of output. Once it has this information, it can be compared with the corresponding level of revenue so as to ascertain the level of profit in each instance. Just as with revenues, it must be emphasized that the assessment of the firm's costs should be a continual one as costs often change reflecting general inflation, or sudden supply shocks to the economy, for example. Thus, when forecasting the future likely costs of a firm the figures are once again obviously estimates and subject to a degree of error. Furthermore, one needs to analyse the behaviour of costs over time, as costs facing the firm in the short run will be different to those in the longer term.

In the short run businesses are faced with a situation where certain factors are fixed. For example, if an entrepreneur purchases, or rents, a small factory, or office, in order to produce a good or service, he or she is committed to this accommodation, at least in the short run, due to the length of the lease. That is, if the firm subsequently wished to increase output beyond the capacity of the present building, it would take time for extensions to be built, or new premises to be found. Therefore, in the short run, only variable factors can be adjusted such as the number of people employed. The long run, however, is a period of time that is sufficient for the entrepreneur to vary all of the firm's inputs so that the optimum size of operation can be reached. The time periods involved between the short run and the long run will depend upon the industry in question. For example, the opening of a new branch office for an estate agent is obviously easier and quicker than the setting up of a new brick works or cement factory, which would involve more significant capital investment.

Due to this distinction between the short run and the long run, each time period is now considered in some depth.

Short-run costs

Fixed costs

As some factors are fixed in the short run, firms will incur fixed costs (FC) that cover these factors regardless of the level of output. These are usually referred to as 'overheads'. Examples of such costs are:

- rent of premises whether a shop, factory or office
- rent of any plant and equipment
- business rates payable to the local authority
- insurance
- the cost of indirect labour not directly involved in production, such as essential management and administrative staff
- charges made for interest upon borrowed monies.

Thus, in the case of house-building firms, whether they build a hundred houses or no houses at all, they would still have to meet these fixed costs. The situation would be the same for a firm of surveyors, for example, whether it undertook any surveys or not. Such costs can only be changed in the long run.

Variable costs

As the name suggests, variable costs (VC) are the costs of inputs that the entrepreneur can alter depending upon the level of output of the firm. Examples of such costs are:

- the wages of productive labour
- the quantities of raw materials or supplies used
- the cost of transport
- the costs of energy used in production.

Therefore, variable costs are dependent upon the level of output of the firm. For example, as the firm builds more houses it will need to employ more staff and purchase more raw materials.

Total costs

From the above analysis it can be seen that in order to produce a good or service firms will be confronted by both fixed and variable costs. These two cost categories added together represent the total costs (TC) facing the firm. This relationship can be expressed in notation form as:

$$TC = FC + VC$$

Where: TC = total costs
FC = fixed costs
VC = variable costs

Average costs

Once total costs have been calculated the average costs (AC) of production can be determined. Average total costs (ATC) are found by dividing the total costs of production by each level of potential output. Both average variable costs (AVC) and average fixed costs (AFC) can also be found in this manner:

$$ATC = \frac{TC}{Q} \qquad AVC = \frac{VC}{Q} \qquad AFC = \frac{FC}{Q}$$

The minimum point on the average total cost curve represents the lowest cost of output given the firm's existing size and short-run fixed factors. However, this is not the point of profit maximization as the level of profit also depends upon corresponding revenues received, as seen from the profit maximizing rule later in this chapter.

Marginal costs

Marginal costs (MC) are the additional costs of producing one more unit of output. Thus, marginal cost can be expressed as:

$$MC = \Delta TC$$

Where: Δ = change in

The behaviour of costs

Figure 8.4 illustrates in diagrammatic form total costs, variable costs and fixed costs and shows how they behave as output changes. From this graph it can be seen that fixed costs remain constant irrespective of output. On the other hand, variable costs, such as the cost of hiring labour, initially tend to increase rapidly as the first few people hired by the firm are likely to be too few to be fully efficient. Therefore, the firm is paying wages in exchange for relatively low output. This situation is represented by the steep gradient OA on the variable cost curve as costs rise at a greater rate than output. After this a stage should be reached whereby additional employees have sufficient numbers of support staff and equipment so that each extra individual employed could work to his or her full potential and thus add greatly to the output of the firm. Therefore, although more wages are being paid, output is increasing at an even greater rate. This situation is represented by the shallow gradient AB on the variable cost curve. This portion of the curve shows great increases in output, yet relatively small increases in the variable cost bill.

However, considering that we have fixed factors in the short run, the law of diminishing returns is likely to eventually set in as output is increased. That is, in order to increase output yet further (beyond Q_B in this example)

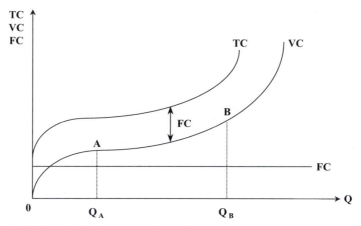

Figure 8.4 *Total costs, fixed costs and variable costs*

overtime payments may have to be made, and more staff may have to be taken on. Under such circumstances wage costs will increase, yet additions to output may be quite marginal as new employees do not have sufficient access to support staff and equipment and therefore their productivity is low. Thus, variable costs now rise rapidly without a corresponding increase in output as shown by the steep gradient of the variable cost curve beyond point B. In a frantic effort to produce even more, under the given short-run conditions, the firm could reach the stage where it had employed so many people that the workplace would become overcrowded and there would not be enough room for the new employees to work even if they wanted to. Here the wage bill would rise without any corresponding increase in output. In fact, at the theoretical extreme, yet more workers could mean that new employees are actually getting in the way and damaging the productivity of existing labour. These problems determine the shape of the variable cost curve. As variable costs are added to the constant fixed costs one finds that the total cost curve is the same shape as the variable cost curve yet at a higher level, the distance between variable costs and total costs being represented by the level of fixed costs.

To illustrate these points yet further, imagine a small firm of house builders that presently occupies a couple of offices in a building. The firm will obviously incur fixed costs, of the type described above, irrespective of its output. That is, in the short run, even in the middle of a recession, if the firm wished to contract in size for example, it would take time to change premises and sell off other equipment such as drawing boards and computers. Alternatively, if the firm was building more during a housing boom it may need to employ more administrative and managerial staff at the head office and thus increase its variable costs. Larger offices may be required to accommodate such expansion, yet if this is not possible in the short run the problem of diminishing returns may well be experienced as pressures on space mean that decisions and work is delayed.

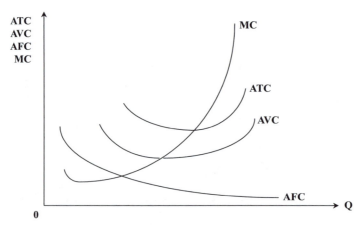

Figure 8.5   *All average costs and marginal costs*

By using logic, or a numerical example, one would find that average total costs, average variable costs and marginal costs, all decrease at first, representing the increasing efficiency of the firm as it expands output, and then begin to increase due to the law of diminishing returns. Average fixed costs will constantly decline as an increasing level of output is divided into a constant cost. These costs have been plotted in Figure 8.5.

Normal profit

The term 'normal profit' is frequently referred to in economics (see Chapter 9). Normal profit is the amount of money that entrepreneurs need to make from a firm in order to keep them in the business. Such an income needs to be at least equal to the opportunity cost of the entrepreneur being involved in the firm. For example, entrepreneurs who have set up their own property consultancy after being employed by another firm, would need a reward at least equal to their old salary in order to keep them in their own business. Thus, such normal profit is considered as a cost of production as it is a necessary payment for the organization of the firm.

The profit maximizing rule

Once information on both costs and revenues has been accumulated one can find out how much profit the firm is expected to make at each level of output. By looking at Figure 8.6 it can be seen that the maximum distance between the total revenue curve and the total cost curve (distance AB) represents the greatest level of profit. For example, the level of profits at Q_1

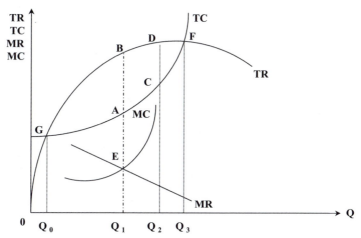

Figure 8.6   *The point of profit maximization*

is obviously greater than that at Q_2 as there is less of a differential between total costs and total revenues at Q_2. That is, distance AB is greater than distance CD. Visually it can be quite difficult to ascertain the exact point of profit maximization in this way, as one has to compare relative distances.

Therefore, it is useful to be aware that the position of maximum profit will coincide with the point where the marginal cost curve and the marginal revenue curve intersect (point E). Thus, in order to maximize profits the firm should produce up to the level where marginal cost equals marginal revenue, or to be more precise, it should produce up to the point where the marginal cost curve cuts the marginal revenue curve from below. This profit maximizing rule can be explained by the fact that for levels of output less than Q_1 yet more than Q_0, the revenue received from producing each additional unit of output (marginal revenue) is greater than the cost of producing each additional unit of output (marginal cost). Therefore, as long as marginal revenue is greater than marginal cost, profits will be increased as more money is being received per unit of production than is being spent to produce and sell it.

However, if the firm produced more than Q_1, marginal revenue would be less than marginal cost for each additional unit of output. As a consequence profits would begin to be eroded away. In fact, if the firm continued to increase production up to Q_3 it would deplete all of its profits, and merely break even, as it sold units for less than cost price. Note that total costs and total revenues are the same at this point (point F). Increasing production beyond Q_3 would mean that the firm's total costs now exceeded its total revenues and an operating loss would be made. It is important to appreciate though that this point of profit maximization (Q_1) could shift over time due to changes on the demand side of the market, as well as possible cost (supply) changes. Being able to forecast such conditions should enable the house builder to decide upon how many houses to build; the building materials supplier to decide upon the most profitable level of output; and the surveying firm to ascertain the most profitable level of work to take on.

The long-run planning decision: the theory of long-run costs

As a business develops, it may feel that the market is sufficiently large to warrant the firm's expansion. Indeed, as growth may take some time such a decision is often based upon a perception of increases in the future size of the market. The firm's existing factory, retail outlet or office may be too small to cope with such an eventuality, and any increase in output in the existing premises, with its fixed factors, will tend to lead to diminishing returns as seen above. Thus, the firm should try to estimate the cost implications of moving into a larger building as part of its long-run planning decision.

In order to achieve this a series of short-run average total cost (SRATC) curves could be drawn up representing the cost structures for different potential sizes that the firm could grow to. In examining the growth of the firm in this way it may be found that various factors, collectively known as economies of scale, tend to cause long-run average total costs (LRATC) to initially decline as the firm becomes larger and produces more. However, logic suggests that if the firm were to become too large, it risks experiencing diseconomies of scale, which are forces that could lead to rising average total costs. Both economies of scale and diseconomies of scale will shortly be discussed at length. As a consequence, given current cost and market conditions, one would find an optimum size or level of output that the firm should aim to grow to. The minimum point on the firm's long-run average total cost curve would identify such a point.

Therefore, forces are present to give us a 'U-shaped' long-run average total cost function. This curve is made up of all the potential sizes of the firm during its growth. As this curve can be estimated prior to its expansion it is sometimes referred to as the long-run planning curve. It should be noted that in the very long run, technological change, for example, might alter the position and level of this curve and the position of its minimum point. For instance, cost-cutting technology could benefit firms of all sizes in the industry and thus lower their short-run average total cost functions. If this were the case, the overall long-run average total cost function, which summarizes these short-run functions, would also be lowered.

Figure 8.7 shows a typical long-run average total cost curve. Here, if only Q_1 were to be produced, the original small firm represented by the short-run average total cost curve $SRATC_1$ would still be the optimum sized firm as it could produce this level of output at the lowest costs of C_1. Note that the larger, 'mid-sized' firm could only produce Q_1 at a higher cost of C_2, perhaps due to the under-utilization of the larger firm's fixed factors. This partially explains why some firms, such as estate agents for example, who need a physical presence in the market itself, have many small branch offices rather than just one large one. However, due to diminishing returns, the

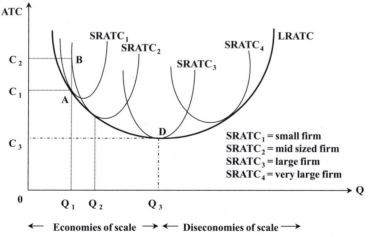

Figure 8.7 *The long-run planning curve*

firm, in its existing state (SRATC$_1$), would find that increasing production only leads to rising costs unless it expanded into larger premises. Therefore, if the small firm wished to produce Q$_2$, for example, it would need to expand into the size of firm as represented by the 'mid-sized' firm (SRATC$_2$). If the business continues to expand, it is likely that it will be able to take advantage of economies of scale so that it could eventually produce Q$_3$ at the low cost of C$_3$, thus being able to undercut the smaller firms.

Economies of scale

Economies of scale can be referred to as economies of size. Essentially they are forces at work that can produce cost savings for firms as they increase in size, expand output and move into larger premises. Many economies of scale exist in both the service and manufacturing sectors. However, some are more applicable to certain firms than others.

Raw material economies of scale

As a manufacturing firm increases its level of production it will require more raw materials to enable it to produce greater quantities of its product. In the same way an expanding firm in the service sector will require more inputs in order to accomplish a higher workload. As a consequence both firms are likely to be able to obtain discounts on the bulk purchase of raw materials and general supplies, thereby decreasing their costs. For example, a large firm of house builders will be a major client for many building supply firms. Thus, in order to retain the house builders' custom, the materials suppliers are likely to offer discounts that directly lower the firm's costs. In addition other benefits such as a better speed of delivery may be offered which can reduce wastage that in turn reduces the builder's costs. Such cost reductions will, by definition, lower the building firms' average total costs. Linked to this idea is the fact that many large firms receive discounts on office supplies such as stationery for the same reason of bulk purchase as demonstrated above.

Financial economies of scale

When financial institutions lend money they do so at their own risk. Lending to a large firm is likely to be seen by the financial institutions as being less risky than lending to a small firm. Such an assessment is based upon the logic that it is more probable that a large firm has considerable assets and possibly financial reserves at its disposal for the lender to draw upon if the firm did not meet its agreed debt repayments. Such increased security enables the financial institutions to offer the loan at a lower rate of interest than would normally be the case if they were lending to a smaller firm. Small firms are unlikely to have the assets or reserves of a larger firm and thus lenders require a higher reward in the form of a higher rate of interest to compensate for the greater degree of risk that they take when lending to smaller enterprises.

However, not only may the cost of a loan be cheaper for the larger firm, but it is also more likely that it will obtain the loan in the first place due to the higher level of security that it can offer the lender. Moreover, there may also be a greater variety of types of finance available to the lower risk large borrower. Such financial economies are an important economy of scale in the construction industry as most projects involve significant sums of money, with projects ranging from large civil engineering work to the extensive refurbishment of existing buildings. Again, such savings will tend to lower the average costs of production of the larger firms.

Research economies of scale

Larger firms are often able to afford expenditure on research and development and normally have their own R&D departments. Conversely small firms are unlikely to be able to devote resources to this area. Research is useful for a variety of reasons. One motive for research is to improve the product so as to generate more revenue through higher levels of demand. However, another important aim of research is to find cheaper and more efficient methods of production, which lower the average costs of production. The house builder, for example, could undertake research in order to investigate ways of producing more houses at a lower cost, subject to conforming to the constraint of acceptable building standards and working practices.

Specialist equipment economies of scale

Larger firms tend to have more money to invest in modern techniques of production. Such techniques tend to be computerized and are more efficient. Therefore, investment in this area is likely to lower the average costs of production. For example, a building surveying practice may be able to complete surveys and give quicker client advice if it has access to the latest laser measuring equipment or has computer-aided design facilities (CAD). In achieving faster surveys the firm could reduce its staffing costs as fewer individuals can undertake and complete more work. Similarly, the large construction firm may be able to purchase a computerized site management system so as to improve productivity on its sites and thus enable them to complete and dispose of their developments at a faster rate. In this instance improvements in productivity could save on costs by reducing the time period over which loan finance is required as well as limiting the duration of the hire of subcontractors and plant.

Managerial economies of scale

As a firm expands, it should be able to spread some of its managerial and administrative costs over a greater volume of output. For example, even if the firm grows threefold it is still only likely to require one manager, one personnel manager, and one accountant, rather than three of each. Thus, the cost of these employees is spread over the larger level of output of the firm.

Despite the potential of economies of scale, there is likely to come a point when the expanding firm will have reached a level of output that places it at the minimum point on its long-run average cost curve given the current state of technology. After this point, diseconomies of scale may set in, and therefore it would not be worth increasing output yet further. This long-run position is represented in Figure 8.7 by the 'large firm's' short-run average total cost curve ($SRATC_3$) giving an efficient level of output of Q_3 at a cost of C_3.

Diseconomies of scale

The term 'diseconomies of scale' encompasses a wide variety of factors that tend to put upward pressure on average costs if a firm becomes too large. Thus, once diseconomies take place the long-run average total cost function of that firm rises. Therefore, diseconomies of scale can be depicted by the rising portion of the long-run average total cost curve as shown in Figure 8.7 as the firm produces an output in excess of Q_3. Many potential diseconomies exist and some of the main ones are discussed below.

Managerial diseconomies

Research into very large firms has revealed that inefficient management is one of the main sources of diseconomies of scale. In many cases communication has proved to become progressively more difficult as firms increase in size. Because of communication difficulties amongst the firm's personnel the risk of mistakes, uncertainty and inefficiency are likely to increase. For example if a firm of house builders develops rapidly from a small local firm into a large national one, it may not be able to cope with the difficulties of large-scale management, and the co-ordination and organiza-tion of many different development sites in different geographical areas.

Specific labour shortage diseconomies

Labour shortages are most likely to occur with businesses that require skilled labour with a specific training. A firm may experience recruitment difficulties as it expands, as it requires more staff, such as managers, effectively to organize the increased level of output. In order to attract this calibre of personnel the firm can offer incentives, normally monetary, such as higher wages or an attractive relocation package. Even if financial inducements are not offered, many firms will pay for the training of some of their staff so that they can obtain the desired skills and necessary qualifications. All of these scenarios will push up the firm's average total costs.

For example, during a property boom that creates an increased level of activity in the built environment, a firm of surveyors may need to tempt new graduates with high starting salaries and other benefits such as a company car. Similarly, an increase in the level of activity in the housing market creates a growth in the rate of new house building. With more developments underway shortages of skilled tradesmen such as plumbers, electricians and carpenters are likely. In these situations house builders have had to pay increased wages and bonuses to such workers in order to attract them to work on their site. The building firm is forced into this situation because if it did not offer enhanced rewards to such workers it would not attract the people it required and therefore the completion of developments may be delayed. This could make the firm miss out on the boom, as it could not sell a completed development at the high prices of the boom.

Raw material diseconomies

The existing sources of materials that a firm draws upon may not be adequate to meet the increased demand for them as the firm expands. Therefore, the firm has to develop new sources of materials, or begin to import them from further afield. Both of these solutions are likely to lead to rising costs. For example, imagine a materials supplier to the construction industry such as a brick manufacturer. If this firm has sited its plant near a suitable type of clay for making bricks it will incur low costs of transport in getting the clay to the factory. However, as the original resource runs out, more expensive, deeper extraction will be required, or clay will have to be transported to the plant from further afield, thus increasing costs.

Market diseconomies

A firm's medium-term growth may be constrained by the size of its potential market. As the firm expands, new markets will have to be sought which could necessitate a considerable investment in marketing. For example, imagine a firm of surveyors that decides to diversify into a new product area such as estate agency. At the very least, money will have to be spent by the firm upon advertising so as to make a name for itself in its new field of operation.

Conclusions on economies and diseconomies of scale

As can be seen, a variety of circumstances can occur that will put pressure on long-run average total costs to initially decrease as the firm expands, but then increase as the firm becomes too large. However several points should be noted:

● The above is by no means an exhaustive list. It simply highlights some of the main economies and diseconomies of scale found in most empirical findings of the growth of businesses.

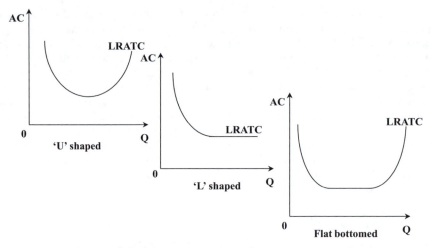

Figure 8.8   *Variations of the long-run average total cost curve*

- Some of the forces mentioned above are only applicable in full to certain types of firm and industry. However, adaptations or approximations of most of these forces are found in the majority of cases.

Some firms will be better than others at exploiting economies of scale and avoiding diseconomies of scale. Such firms are likely to face 'L-shaped' or 'flat-bottomed' long-run average total cost curves, rather than the 'U-shaped' curve that the traditional theory predicts. These alternative cost patterns can be seen in Figure 8.8 with the last two curves showing the successful avoidance or postponement of any diseconomies of scale.

Note that firms often grow in size by merging with others. A horizontal merger is where a firm merges with another firm in the same line of business. An example would be the merging of two financial institutions. A vertical merger, on the other hand, is where a firm merges with others who are in related areas but not the same one. For example, a firm of surveyors may merge with an estate agency or building society. Finally, one can have conglomerate mergers whereby firms in unrelated areas of business join together.

9 Market structures in construction and property

A knowledge of the fundamental economic forces acting upon any firm, or group of firms, in any industry, helps one to understand an almost limitless range of applied issues such as:

- how a firm's costs may change as it develops
- the implications for a firm of changing levels of demand
- how profitable firms are, or are likely to be
- how a firm's profits can be enhanced
- how much a firm produces
- how many people are, or will be, employed
- explaining regional variations in the prosperity of firms in the same industry
- how general economic conditions, such as a recession, can affect the firm
- how both local and national government policies can affect the firm.

This list is certainly not an exhaustive one as the theories of the firm that now follow are exceptionally versatile and can therefore address many other points.

As a starting point, this chapter identifies the main types of industrial structure that are normally observed in most economies. It should be appreciated that when attempting to categorize industries, or indeed parts of industries one will normally find a spectrum ranging from firms operating in a highly competitive environment, to those who face little, or no, competition. The exact situation that the firm finds itself in will affect all of the points listed above. To help in this task of classification, economics has devised specific terminology to identify particular types of industrial structure as can be seen in Figure 9.1. This diagram represents a spectrum of industrial structure from that of perfect competition to that of monopoly. Specifically it identifies the following possibilities:

- Perfect competition. This is where a large number of fully competitive firms in the industry produce identical goods or services.
- Monopolistic competition. In such markets there is a large number of firms in the industry who are able to reduce the level of competition by differentiating their good or service in some way so that their product has a specific, individual identity.

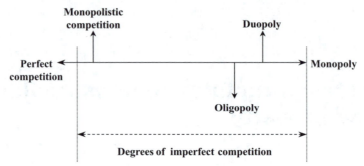

Figure 9.1   *The spectrum of industrial structure*

- Oligopoly. This exists where several large firms dominate the industry. If these firms collude together to agree upon pricing and output, a cartel is said to be formed. Cartels are often thought not to be in the consumers' interests and are outlawed in many countries and in the field of international trade.
- Duopoly. This is where two large firms dominate the industry.
- Monopoly. A true monopoly will occur if there is only one firm in the industry and there is therefore no competition.

Thus, it can be seen from the above that there are two analytical extremes: very high (perfect) competition on the one side, and no competition (pure monopoly) on the other. Any industry whose characteristics are different to these will fall somewhere in the middle of this spectrum. It is important to note that all forms of market structure, except pure perfect competition, can be categorized under the overall heading of imperfect competition.

In reality, however, it is unlikely that one will see an industry that is perfectly competitive, or a firm that has a complete monopoly. The study of these situations enables one to understand the behaviour of, and the implications of, firms that approach either extreme. By examining these structures one should be able to form an opinion as to whether increased competition is beneficial to the consumer or not. Conversely one should also be able to assess the implications of a decrease in competition in any market.

The theory of perfect competition and its application

Many parts of the property industry are characterized by firms working in such highly competitive markets that they approximate the conditions laid out in the theory of perfect competition. It must be realized, however, that although this statement is true in many instances, it is at risk of being a generalization. For example, some small, perhaps remote, communities may only be served by one or two local firms thereby reducing the degree of choice for the consumer. In such instances if repair work is needed on a house, for example, the choice of suitable firms can be severely limited. In other words, data may show that there are thousands of firms involved in building repair work, yet it is the distribution and accessibility of these firms that is important. What is, however, a more realistic stance is to firstly select an area where one knows that a degree of competition exists. For purposes of illustration, if examining a number of house builders developing sites in a region one could try to assess the degree of competition between them, and how well their behaviour approximates the model of perfect competition. If the structure of their industry does approach the conditions of the model, the theory can then be used to draw some useful conclusions and answer such points as those raised at the beginning of this chapter.

Thus, as implied above, to find an industry that truly reflected the model of perfect competition would be nearly impossible as such a large number of restrictive criteria need to hold in order to produce this perfectly competitive state. However, it is true to say that some industries are so highly competitive that they do indeed approximate the theory. It should also be noted that it has been the policy of many governments, primarily on the grounds of increasing national efficiency, to encourage a higher rate of competition between firms. In fact, some politicians have used the theory as the goal of a perfect market and a liberalized economy. Therefore, by examining this model one can begin to see the rationale for, and the implications of, such public objectives.

There now follows an explanation of the conditions or assumptions of the theory of perfect competition along with a debate regarding their realism and application to real-world scenarios in the general property arena.

Condition 1: a large number of firms

For the theory to hold there must be a situation where there are such a large number of firms in competition with one another, that any one firm only produces an insignificant share of total industry output. Thus, any change in output by one firm has little or no effect upon the overall market and the ruling market price. As such, because the firm has to accept the going market price for its good, and cannot influence it in any way, the firm is known as a price taker. Diagrammatically this can be seen in Figure 9.2. Here it is shown that even if the individual firm changed its output from q_1 to q_2, the changes would be so small in terms of overall market supply that they would not influence the market equilibrium price of P^*, or equilibrium output Q^*. In other words the individual firm's contribution to the market share is so small that changes in its output are hardly noticeable.

As the firm cannot influence price it must, by definition, be faced with a perfectly elastic demand curve. Moreover, under this model, there is no incentive for the firm to deviate from this market price by overcharging or undercharging for its output. For example, if it accepts the market price of P^* and produces q_1, its total revenue would be equal to the area OP^*Aq_1. If, however, it attempted to push prices above the market equilibrium, it would potentially lose all of its customers as they could buy the same good or

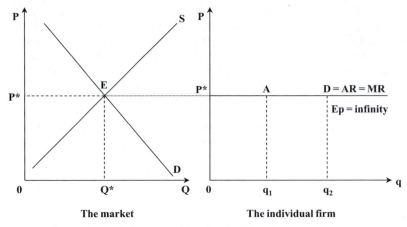

Figure 9.2   *The market and the competitive firm's demand curve*

service from one of the many other firms in the industry at the cheaper ruling market price. Moreover, if the firm charged less than P*, perhaps in the hope of attracting more clients, this would simply lead to a decline in its total revenue as the firm can sell all that it wants to at the market price anyway.

Applying this to the example of a firm of house builders, one can take the realistic assumption that there are a number of different house-building firms developing sites in and around a particular town. These developments are likely to contain a highly similar mix of houses so as to conform to planning requirements and current consumer tastes. Thus, the market price for new houses will be set in line with the choice that is available to the consumer, and the available supply from these firms. It should be noted that the firms are also in competition with the vast number of second-hand houses that are available on the market through estate agents, or privately arranged sales advertised in newspapers. Thus, if any one building firm tried to increase its prices over the market price, it is likely to lose sales to its competitors as consumers can buy similar houses on a different development. Also, there would be little point in the firm undercutting the going price for a particular type of house if there are a sufficient number of buyers to buy all the houses offered for sale on the market. Such price cutting behaviour would only mean a loss in potential revenues. Therefore, it should be apparent that an element of market research is crucial for the building firm to ensure an effective and successful pricing policy. However, the builder would have to recognize that the market price itself is subject to continual change. Such price variations would primarily be due to changes in the overall level of market demand for such dwellings. Note that with the theoretical extreme of an inelastic demand curve, average revenue is constant and the same as marginal revenue as the firm can sell as much as it wants without having to alter the price.

Condition 2: product homogeneity

As already suggested, for the theory of perfect competition to work in its purest form, it must also be assumed that all firms within the industry are producing an identical product. This second assumption, combined with the first, ensures that the firm has no control over price. That is, the firm is also unable to change the price by differentiating its product to give it superior characteristics for example. This assumption may, at first, sound absurd. However, one can see how this assumption can be applied to the 'case study' of house builders in a town.

First, one could argue that builders could, and do, differentiate their product. Although one observes many house builders producing, say, three-bedroom semi-detached houses, different building firms may use different designs, different qualities of materials, different standard of finish, landscaping, and so on. In fact, location itself, even within a town, could differentiate one housing estate from another. For example, whether the houses are near an industrial estate, or overlook pleasant views of the countryside or sea, will be key factors in determining the attractiveness of the houses to the potential buyer. Moreover, some building firms may seek to differentiate their product by financial means such as offering special discounts to first-time buyers, or offering subsidized or 'low start' mortgages. Obviously advertising is a key component of such a strategy.

However, consumers who are keen just to live in the town, perhaps due to the proximity to their workplace, may be unaware of some of the differences mentioned above. For instance, build quality can be covered up by the finish, and many would not know the difference between a well-built house and one which has not been so well built, simply because few have an intricate knowledge of the building process itself. Furthermore, if building firms are in competition with one another, they are likely to copy each other's innovations so that their developments look equally attractive to clients. If one building firm used extensive landscaping as a positive selling point for their estate, other builders may be forced to follow suit if they found that their market share was declining. Thus, the assumption of product homogeneity may not be as far-fetched as originally perceived. In the service sector, the condition is far easier to perceive. A structural survey from one firm of surveyors should be highly similar to the survey of another firm.

Condition 3: perfect ease of entry into and exit from the industry

This third assumption means that no barriers to entry or exit exist in the industry. Therefore, if a firm wishes to set up in the industry it can freely do so, just as it could leave the industry if it so desired. This is an important assumption as it ensures that, in the long run, all firms can only make normal profit. Normal profit is a level of return which provides just enough reward to keep the entrepreneur in business as it is where total revenue equals total costs, where payments to the entrepreneur are included in total costs. Whilst firms are only making normal profit there is no pressure on

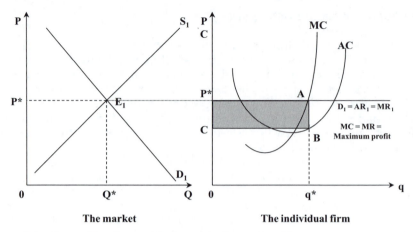

Figure 9.3   *Competitive firms with short-run abnormal profits*

firms to enter or leave the industry. However, short-run changes to this situation can occur due to market-led disturbances creating either short-run profits or short-run losses. Examples of such changes, and their implications, are discussed below.

To understand the importance of this assumption, imagine that existing firms in the house-building industry are initially making excess or super-normal profits, i.e. profits in excess of normal profits. (Please note that super-normal profits can also be referred to as abnormal profits.) This situation could have been brought about by a sudden rise in demand pushing up revenues, or by the introduction of cost-cutting technology that lowered the cost of production. The existence of super-normal profits is shown in Figure 9.3. Here, each individual firm is receiving the market price of P* and producing q* output. Because of their cost structure relative to their revenues, they are making super-normal profits equal to the area CP*AB. If super-normal profits could be earned in the industry, others would be attracted to join that industry or existing building firms may direct more resources into the house-building sector, so as to take advantage of the high level of profits. As more and more firms enter the industry, or direct resources into it, total industry supply (the market supply of new dwellings in the town) would begin to shift to the right. In fact, this process would continue as long as super-normal profits could be made and planning permission was granted for such additional developments.

However, as industry supply increases, the levels of these additional profits will be driven down as the market price is driven down due to increased competition and subsequent consumer choice. Thus the market will eventually reach the stage where only normal profits are being earned and as a consequence no further firms will be attracted into the industry. Such a sequence of events can be seen in Figure 9.4. This diagram shows that total industry supply has increased from S_1 to S_2 leading to an increase in the industry's overall output from Q* to Q_2. As

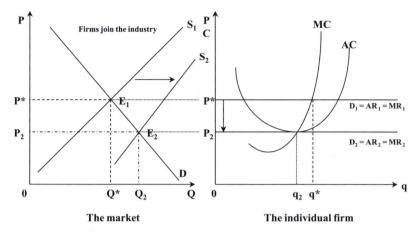

Figure 9.4 *Competitive firms and the long-run reattainment of normal profit*

more is now available to choose from, prices have been driven down from the original price of P* to the new, lower price of P_2. Note also that due to increased competition amongst the firms in the industry, each firm's individual market share has been reduced from q* to q_2.

So as to understand the sequence of events that are likely to occur if a loss were being made by the industry, imagine that highly competitive house-building firms are initially earning normal profits. In Figure 9.5 normal profits are being achieved where the market price is given by P_1. Such a price would be generated by the interaction of the demand curve D_1 and the supply curve S_1. It as at this point where firms are individually producing a level of output equal to q_1. If the firms are then faced with the onset of an economic recession the demand for housing is likely to fall. As the demand for housing decreases, losses will begin to be made by firms in the industry. That is, building costs remain the same, yet the market demand for houses falls from D_1 to D_2, thereby lowering the market price to P_2. Under such circumstances, the output of individual firms drops from q_1 to q_2 as the demand facing each firm falls to D_2. The short-run result is that the firms incur losses equal to the area P_2ABC. Firms would lower output to this new level, as it is here where marginal costs are again equated with marginal revenue. Although this point of intersection normally gives a position of maximum profit it equally advises the entrepreneur as to the location of the least loss. Faced by such adverse circumstances, some firms may not be able to sustain such losses in the long run and are likely to go out of business.

As firms fail and leave the industry, industry supply will be driven back from S_1 to S_2. This contraction in supply will reduce output and therefore choice, which in turn will put upwards pressure on price until the market is again in a situation whereby the remaining firms are making normal profit. The end result is that the price of P_1 is re-

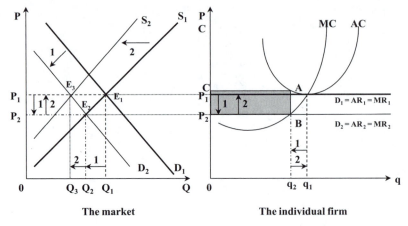

The market   **The individual firm**

Figure 9.5   *Competitive firms making short-run losses*

established although the industry has become leaner as fewer surviving firms means that industry output falls to a level of only Q_3 rather than Q_1. Therefore, in conclusion, declining market demand is likely to lead to lower house prices with the weaker building firms going out of business. However, as more firms go out of business, new houses can become short in supply thus driving prices higher again as people compete for the limited number of dwellings. Therefore, in the long run we are back in the situation where all (surviving) firms are making normal profit. However, in reality, building firms could at least attempt to prevent themselves from going bankrupt in the following ways:

- Attempt to reduce their cost structure. By looking at Figure 9.5, it can be seen that if costs were lowered, losses could be reduced or eradicated. Firms could seek to lower their costs by laying off non-essential staff, enhancing productivity with their existing labour and capital, reducing site wastage and theft, using cheaper materials, constructing simpler designs, and so on.
- Try to increase the level of demand facing the firm via the methods of potential differentiation mentioned in the critique of condition 2 of the model (see above).
- Strive to increase the level of demand facing the industry as a whole. For example, house-building firms can put joint pressure on the government to lower the interest rate, thus lowering the cost of borrowing for the purposes of mortgages for house purchase. In fact, the lowering of interest rates will have the dual impact of not only stimulating consumer demand, but also lowering development costs for the building firms as the cost of borrowed funds in general declines. Alternatively, firms in the industry may join together to finance an advertising campaign trying to persuade consumers of the merits of buying a new home.

Condition 4: perfect knowledge on behalf of both producers and consumers

For a truly competitive environment to exist both producers and consumers need to have a very good, or near perfect, knowledge of all aspects of the market. In the case of producers, as already suggested above in part of the discussion in condition 2, the model assumes that all have a perfect knowledge of existing and new techniques of production. For example, if one firm were to have a technological breakthrough, all other firms would be able to copy it. In other words there can be no lasting trade secrets amongst firms in the industry. With the example of the house-building firm, information about innovations could be gathered through trade magazines, the observation of competitors' sites, plus information passed on by subcontractors working for more than one firm. In the same way a firm of surveyors would quickly learn of new techniques used by a competitor such as an increased use of computer-aided design. As a consequence no one firm can gain technologically induced cost advantages over another. Therefore, all producers should be equally efficient.

For competition to work consumers also need to have a perfect knowledge of ruling market conditions such as price, availability and quality. Such information would ensure that consumers would not purchase a good or service for a price in excess of the market price, as they would know that they could obtain it at a more reasonable price elsewhere. With respect to buying a house, the potential purchaser through visiting local estate agents, seeking the advice of a surveyor, or reading the property section of relevant local newspapers could gain this knowledge. Moreover, as housing is normally an expensive commodity in relation to income, it is highly likely that people will take extreme care in the purchasing process. Due to this fact, people are likely to visit many houses before making a final decision. Thus, consumers should obtain a relatively good knowledge of local property prices. In any case, when an application to borrow money for a house purchase is made, the lending institution would formally send out a surveyor to realistically value the property given current market conditions. In this way the consumer should obtain an accurate professional judgement of the ruling market price independently of any prior assessment of price. These processes should enable the house buyer to offer the correct market price for the house.

In the case of older properties, the consumer could also instruct a building surveyor to undertake a full structural survey of the dwelling. A structural survey may indicate the need for money to be spent on remedial work, such as the need to combat damp, or replace roof tiles for example. On receipt of the surveyor's advice the purchaser could attempt to reduce his or her offer below the market price so as to make an allowance for expenses that may have to be incurred either at some future date or immediately in order to satisfy the mortgage lender.

Condition 5: identical factor prices

The term 'factor prices' refers to the prices that firms have to pay for the various factors of production required to produce their good or service.

Having identical factor prices implies that all firms are confronted by highly similar cost conditions. That is, the cost of items such as raw materials, labour and plant hire is the same for all firms in the industry. Such an assumption is quite likely in the house-building market, at least at the local level, assuming (as the model does) that all firms are roughly the same size. Indeed local surveying firms could be similarly categorized. However, it should be noted that if there are larger firms in the market they are likely to benefit from cost advantages brought about by economies of scale (see Chapter 8). In practice such a distortion of the theory caused by the existence of a large firm is not critical to the general workings of the model. The reasoning behind this statement is that on average there will still normally be a very large number of firms in competition, and the majority of these firms will be facing similar cost conditions. In other words the theory may not work completely perfectly, but it can be adapted to examine the majority of competitive firms under consideration.

Condition 6: a large number of consumers

Another prerequisite for a highly competitive environment is that it will only exist where there are so many buyers in the market that no one buyer purchases a significant proportion of output. As a consequence, if one consumer altered its demand, for the good or service in question, it would have such a small impact upon total demand that market prices would not be affected. This condition is relatively easy to uphold in the case of the housing market, as few will buy houses in bulk. This is not to say that distortions do not exist. House prices can be driven up, for example, if a large firm moves to a town and decides to accommodate its personnel in company houses. In instances like this one company may be the sole purchaser of a large amount of housing stock. However, in relation to the total size of the housing market such activity is still likely to be comparatively rare and insignificant.

Conclusions on the model of perfect competition

In reality, it is unlikely that all of the conditions of the model of perfect competition will hold and therefore one will perhaps never observe a perfectly competitive market in practice. However, there are examples of situations where firms operate in markets that exhibit such a high degree of competition that this model can be useful to demonstrate the likely behaviour of these firms. Moreover, the model is used as a 'yardstick' to enable analysts to see the potential advantages and disadvantages of a perfectly competitive industry. Thus, a government that wishes to encourage increased competition in the economy, for example, could examine the model to see the implications of such a policy. Throughout this debate, we have considered the situation of many small firms of house builders competing for their market share in a town. It has also been noted that the analysis may become distorted, although not necessarily nullified, if a large national firm of house builders becomes involved in the market. However, the nearer one gets to this model, the easier it becomes to perceive a situation whereby there will be a large number of similar houses being built by different firms, selling at a price that is primarily determined by the degree of competition amongst firms and the level of consumer demand.

The theory of monopoly and its application

Before discussing the implications of monopolies on the construction industry, it is important to note that, in everyday speech, the term 'monopoly' is frequently used inaccurately as people are actually referring to situations where a few firms dominate the industry rather than just one. Having said that, monopolies and 'near monopolies' do exist and the text now moves on to examine the relative disadvantages and advantages of such a market structure.

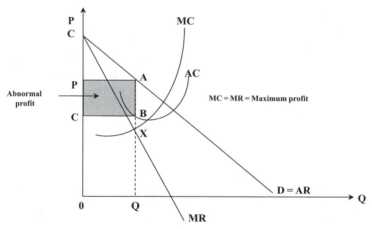

Figure 9.6 *The cost and revenue structure of a monopoly making abnormal profit*

A pure monopoly is the extreme situation whereby the firm *is* the industry, i.e. there is only one firm operating in that industry. Therefore, the firm's demand curve will be the market's demand curve, and for the usual reasons one would expect such a demand curve to be downward sloping (see Chapter 1). Thus the cost and revenue structure of a monopoly will be like that shown in Figure 9.6. This diagram indicates that the monopoly firm will strive to produce an output of Q, and sell it at a price of P. Just as with the theory of perfect competition, the profit maximizing rule states that the firm will maximize its profits where its marginal cost curve cuts its marginal revenue curve from below as can be seen at point X. As with this example, where price (average revenue) exceeds average costs at this profit maximizing level of output, it is normally assumed that monopolies make

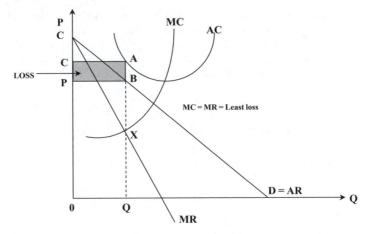

Figure 9.7   *The cost and revenue structure of a monopoly making a loss*

super-normal (abnormal) profits. Such abnormal profits are over and above the profit necessary to keep the entrepreneur in the industry and are represented by the shaded area PCAB.

However, the existence of high levels of profit is not always guaranteed for firms in a monopoly situation as this will depend upon the level of demand for their good or service, and their ruling cost structure. For example, if operating costs were to rise significantly, say due to escalating energy costs, a firm could face a loss-making situation until it managed successfully to react to the problem, or go out of business. A drop in demand could bring about a similar situation. This could occur if some consumers chose to boycott the firm's products if they believed them to be environmentally unfriendly, or perceived that the firm had operations in a politically unpopular country. At the extreme this could mean that the firm were to suffer losses. Such a loss-making situation is demonstrated in Figure 9.7. Here, at an output of Q the average costs of production, C, are higher than the selling price P. As such, a loss is being made equal to the area PCAB. However, at this level of output, given the current conditions, losses are being minimized as the firm is operating at its optimal level of output whereby marginal costs and marginal revenue are equated. In other words, at any other level of output losses would be even higher.

In some countries, government-owned monopolies may actually make long-run losses as other policy objectives, other than profit maximization, may be deemed to be more important by the public sector. Such long-run losses can be financed from taxation revenues, or from cross-subsidization from another profit-making part of the public sector. Alternative goals that the government may wish to achieve from industry are varied. For example, they may wish to keep a loss-making industry open so as to secure employment in a depressed region of the country. Or they may wish to ensure that the country can produce a good of strategic importance, such as defence equipment, so that they do not have to rely on imports at a time of crisis. As an additional applied example, one can imagine the possible situation whereby a country is unable to acquire cheap imports of a key building material such as cement. Under such circumstances the government may conclude that producing domestically at a loss is the only way of obtaining such an important product in order to facilitate adequate levels of construction in the economy. The domestic industry is likely to incur losses in the first few years of production as the cost of setting up, and providing the necessary capital equipment would have to be covered.

Alternatively, even if profits were being made, the government may wish that more is produced and at a lower price. Thus, the firm could be instructed to produce beyond the profit maximizing level as shown in Figure 9.6, by producing a level of output in excess of Q and at a price less than P. Such action may be felt to be necessary to ensure that an essential service such as public transport is provided in all areas at an affordable price to all members of the community. Whether one examines the scenario of the monopoly making a loss, or suffering reduced profits due to direct government intervention in relation to pricing and output, the industry is likely, at least in the former case, to come under public ownership. That is, such industries would not attract the private monopolist, as profits could not be made. Therefore, to ensure the ongoing existence of the industry in such circumstances the firm would need to be nationalized.

However, if the monopoly firm does make super-normal profit, in the absence of anti-monopoly legislation or control, it can maintain such profits even in the long run. This long-run retention of high profits is possible as the existing monopolist can prevent potential competitors from joining the market by taking advantage of the existence of barriers to entry. This feature is in direct contrast with the theory of perfect competition where only normal profits can be made due to the allowance of freedom of entry into the industry. In the case of a monopoly, entry barriers can be found such as:

- The firm may have a unique patent or licence to produce, or provide, a particular good or service that it has invented and developed.
- The monopoly may own and therefore control all of the raw materials necessary for production.
- Specific economies of scale could dictate that there is room for only one efficient firm in the area (in the case of a local monopoly), or in the country (as in the case of a broader, national monopoly).
- The monopoly may be created by, or at least protected by, the government. This may be done for a variety or reasons as discussed above. A further example could be the preventing of the emergence of a whole range of electricity firms in an area. The rationale for stopping competition in this case would be the fear that having a number of firms may require buildings to be wired up to several different firms so as to give the present, or future, occupier the choice of service. Therefore, government action preventing competition may avoid the wasteful duplication of scarce resources.

Such barriers to entry can give the firm significant monopoly power. Monopoly power can be further enhanced by the extent to which the monopoly can control the consumer's purchasing decisions. This latter point primarily depends upon the availability of substitutes for the firm's good or service. For example, a monopoly in wooden window frames is unlikely to be a very powerful one, as many close substitutes to wooden window frames, namely plastic and metal, do exist. However, a monopoly in the provision of scaffolding, or cement, may be more powerful. Thus, monopoly power not only depends upon the control of supply, but also upon the elasticity of demand, whereby the more inelastic the demand curve facing the firm, the more powerful it can be. Monopoly power is an important consideration in the construction industry as it is often found that monopolies, or at least situations close to monopoly, dominate the materials supply side of the building industry in many countries. Such a market structure can occur in the materials supply industry for a number of reasons:

- Many building materials and components, such as cement for example, are standard and subsequently relatively homogeneous. In such a situation there is considerable scope for significant economies of scale to be gained by one large firm. Any firm that benefited from such economies could produce at such a low cost that it could aggressively price potential competitors out of the market.
- Many construction materials, such as aggregates for example, are heavy and bulky in relation to their value. This feature, coupled with high transport costs, rationalizes the case for local monopolies.
- Some firms have managed to protect their monopoly position by creating a high degree of brand loyalty for their product. Such brand loyalty may be achieved in a variety of ways, such as through prompt deliveries and back-up services, the maintenance and standardization of quality, the provision of short-term credit, etc.
- Some suppliers may also be the owners of the particular resource. For example a monopoly timber merchant may also, perhaps via vertical integration and the acquisition of another firm, be the owners of the forests from which the wood comes from in the first instance.
- Cyclical fluctuations in the prosperity of the construction industry inevitably have a knock-on effect upon the materials supply industry. Thus the materials supply industry must be structured in such a way as to accommodate variable demand. Whereas construction firms normally have the advantage of a low capital base (due to the well-established hire sector that exists for capital equipment for construction in most countries), and a flexible employment structure (due to the hiring of subcontracted labour rather than full-time employees), this is not the case in the materials supply industry. Here, many of the industries require a large amount of plant and machinery and such capital intensity means that materials supply firms are not as well prepared as building firms to cope with short-run changes in output. In order to survive under such circumstances the situation encourages larger firms, thus strengthening the trend towards concentration in the industry.

Arguments against monopolies

There is frequently a general assumption that monopolies work against the consumer's interests. In the example below, remember that it is the materials supplier who is often close to being a monopoly, and it is the construction industry that is the 'consumer' in this case. However, it is also important to note that some very large construction firms themselves may be virtual monopolies in specialized markets such as offshore work, tunneling, road building and other key areas of civil engineering. The text now goes on to describe some of the main arguments against monopolies.

Due to their control over the market, monopolies can reduce output and thus force prices upwards as the good or service becomes shorter in supply than would be the case if there were to be higher levels of competition in the industry. In other words, it is argued that less is produced and at a higher price than would be the case if the industry was perfectly competitive. Note that as the demand curve is determined by the consumers, a monopolist can change either price or quantity and not both. In Figure 9.6, for example, it can be seen that if the firm limits output to Q it can charge the high price of P. However, if it produced anything in excess of Q it would have to lower prices, as otherwise it would accumulate unwanted stock.

For purposes of comparison between two market structures it is shown in Figure 9.8 that a monopoly will tend to produce at a high price (Pm) and a low quantity (Qm) as this reflects the profit maximizing point where its marginal cost curve cuts its marginal revenue curve from below. However, the perfectly competitive firm will tend to produce more (Qp) at a lower price (Pp). At this level of production the perfectly competitive firm will also maximize its profits as with such a market structure the marginal revenue curve and the average curve are one and the same thing (see Figure 9.2). Thus, as profits are at their greatest where marginal cost and marginal revenue are equated a monopolist will achieve this at Qm (see point X on Figure 9.8) and a perfectly

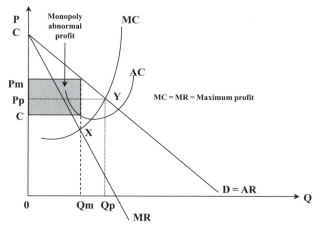

Figure 9.8 *Monopoly and perfect competition compared*

competitive firm will do so at Qp (see point Y on Figure 9.8). (See also the discussion on the profit maximizing rule in Chapter 8.) Therefore, if monopolies dominate the materials supply industry, construction firms may be faced with a poor choice of components at an inflated price.

As barriers to entry into the industry are likely to exist the monopolist may have less incentive to improve the quality of the product, as the firm knows that customers are unable to purchase the good or service elsewhere. The degree to which the monopolist can rely upon this market characteristic will depend upon the degree of monopoly power that the firm enjoys. Moreover, if this argument holds, the logical extension of this analysis is that the monopolist could also become increasingly complacent with respect to maintaining the current quality of its output and the punctuality of its delivery times, for example. As a consequence, far from the normal expectation of a product or service improving over time, it could in fact become increasingly defective. Therefore, construction firms may have to put up with second-rate materials that are not delivered on time or in the correct quantities. Moreover, they may be unsure of the quality of different consignments of supplies. Such difficulties could add to the time of construction and thus increase building costs. If such cost increases cannot be passed on to the consumer, due to depressed demand for example, the profitability and survival of some construction firms could be at risk.

Continued super-normal profits will generate vast resources of financial capital that the monopolist could accumulate and then use to increase its overall monopoly power yet further. This could be achieved, for example, by horizontal integration (the buying up of competitors in the same line of business), by vertical integration (the buying up of suppliers and/or sales outlets for the product), or by a conglomerate merger (the buying up of firms not related to the monopoly's original product base). Such merger activity may compound the problems facing building firms.

For these reasons, governments have shown concern about monopolies in the private sector, and have often legislated to prevent them being created in the first instance. In addition they have sought to place controls on those that already exist. In order to achieve this, a government body such as a monopolies and mergers commission may be set up to monitor the degree of competition between firms in an industry and its likely impact upon the consumer. However, government and consumers alike may be heartened by the fact that the power of a monopoly can be eroded automatically over time by such factors as:

- Import penetration. If domestic monopolies push up their prices too far, or let quality slip, it will increase the attractiveness and feasibility of firms using foreign suppliers. However, the practicality of such import penetration may be limited due to the low cost, high weight and volume nature of many building supplies such as bricks and cement.
- The availability of near substitutes. As more alternatives come on to the market, consumers would no longer need to rely exclusively on the monopoly's output. For example, a firm that enjoyed a monopoly in the supply of wooden window frames may have to become more competitive as more modern plastic window frames come on to the market.

- The power of the consumers themselves. The consumers of the monopoly's output could consist of very large firms. As such one consumer could account for a large share of the monopoly's business. Thus, if the materials supply monopolist failed to provide an adequate service at an acceptable price its behaviour could impair the performance of the construction firm that it supplies. If the construction firm failed, or got into financial difficulty, it would reduce its demand for materials that would in turn damage the monopoly supplier. Furthermore, poor quality and inflated prices would encourage firms to seek alternative materials or methods of construction so as to remove the reliance upon the original monopoly.

It should also be noted that there are also some strong arguments in favour of monopolies, and these are discussed below.

Arguments in favour of monopolies

Although there are many logical arguments against monopolies there are also some potentially positive features of this market structure.

The fear that a monopoly has a lack of incentive to maintain or improve its product may be unfounded. Rather, large firms could actually be more likely to be able to finance expensive research and development via their ability to fund such activity through the accumulation of super-normal profits. Such research expenditure could have two major benefits for the end-user of the firm's product or products, as follows.

1 The existing product or product range of the monopoly could be updated and improved. This should ensure that the construction industry could obtain modern, state-of-the-art materials and that their incorporation could be used as a selling point for their completed buildings. For example, the developer could claim that its office block contains the most modern and efficient services, or that it has a very high energy-saving rating due to a new, improved form of insulating material used in the building's construction.
2 Technological improvements could make the monopoly's output cheaper to produce. If this were to be the case, the supply curve should shift to the right (see Chapter 1) thereby lowering prices to the consumer.

As suggested above it would be unwise for a monopoly supplier to behave in such a manner that could jeopardize the very existence of the firms that it supplies. Obviously the survival of both the supplier and the end-user are highly interrelated. For example, if a materials supplier were to increase prices too much many construction firms would go out of business. If this were to happen the level of new build is likely to drop and demand would have to be accommodated via more extensive use of existing buildings, change of use, or refurbishment. None of these alternatives would provide the monopoly with the same level of demand for its output, as would be the case for new build.

The accusation that a monopoly could reduce the quality of its output may also be unsound. Government regulations could easily be set, and regularly updated if required, to ensure that construction materials were of an adequate, and perhaps improving, standard.

The majority of the views against monopolies rest upon the assumption that such firms are making super-normal profits. However, research has revealed that many monopolistic firms actually have sales revenue maximization as their goal rather than profit maximization. The explanations behind this view are varied, but one key theme of empirical studies in this area is that as management is normally separate to ownership, they are unlikely to be obsessed with maximizing profit. Instead it is felt that managers could simply produce up to the point where they are ensured sufficient profits for the survival of the firm and for the satisfaction of its shareholders. Alternative views to explain such behaviour are that firms wish to saturate the market for reasons ranging from longer-term profit maximization to mere personal ego.

If, in most cases of monopoly, sales revenue maximization, rather than profit maximization, is observed it suggests that monopolies may well

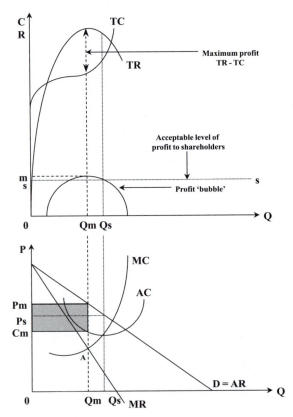

Figure 9.9   *Profit maximization versus sales revenue maximization*

produce more, and at a lower price, than the initial theory would predict. This debate is summarized in Figure 9.9. This diagram shows that if a firm were to maximize its profits it would produce Qm output at a price of Pm as this is where its marginal costs are equal to its marginal revenue, i.e. where there is the maximum distance between its total costs and total revenue. Perhaps the easiest way to see the potential level of profits is by looking at the 'profit bubble' which is simply the area between total costs and total revenues brought down to the horizontal axis. This bubble has the same area as the shaded rectangle. Maximum profit is shown at 'm', which is derived from the peak of the bubble. However if shareholders were satisfied with a profit level shown by point 's' the firm could produce more, in this case Qs, and at a lower price of Ps. Thus, in reality, the level of output that the monopoly firm produces, and the price that it charges, may in fact be far closer to that seen in a competitive market.

Conclusion on the theory of monopoly

Just as with the theory of perfect competition, it must be emphasized that the model of monopoly is a theoretical extreme that is unlikely to be observed in its purest form in the real world. However, many firms do approach such a market structure, and as such this model can give indications of the characteristics of that market. The nearer a firm is to being a monopoly the closer the model will be in providing accurate predictions. Moreover, the model helps government decide upon whether to encourage competition or to promote restrictive practices. The full impact upon the construction industry of having monopolies in the materials supply sector, or indeed the impact upon the consumer of having some parts of the construction industry itself dominated by near monopolies, depends upon the particular firm or firms in question, and the weight of the arguments posed in the analysis above in each instance.

The construction industry

To examine the make-up of the construction industry in a detailed and meaningful way the area of study must first be clearly defined. In other words what is meant by the term 'the construction industry' needs to be identified in order to include all its relevant parts, but to exclude services that, although related to the construction industry, would be more logically associated with another industry. To help with this task many countries have official classifications of industries such as a standard industrial classification (SIC). However, care must be exercised when making use of such data as classifications differ between countries, and definitions of industries have changed over time within countries themselves. Perhaps a useful guideline is to limit the definition of the construction industry to those firms involved in the actual assembly process of new buildings and the repair and maintenance of existing buildings. This would then leave out related, but dissimilar industries such as the materials industry and the provision of services such as surveying and development finance.

The construction industry and the economy

The exact nature of the construction industry will differ from country to country. However, international comparisons do reveal a certain degree of similarity, therefore in most cases the following generalized comments can be made.

Due to the nature of its output, the construction industry is normally very large in most countries and thus accounts for a high proportion of overall national output. If the related industries of the materials supply sector, estate agents, and surveying are added, for example, the significance of construction in the economy is yet further emphasized. One should not be surprised at the magnitude of the industry as one must appreciate that most forms of economic activity require some form of building whether it be an agricultural warehouse, an office, a factory or a retail centre. Moreover, everyone needs a home to live in, and there will also be a substantial building stock in existence that requires repair and maintenance, or replacement due to ongoing obsolescence.

An increasing trend has been the globalization of the construction industry, and its related industries. It has been observed that there has been an increase in firms competing for projects abroad as well as on the home market. As such, many countries have a plethora of multinational construction firms involved in the development process. Similarly, there has been a marked growth in the merging of firms on an international basis as one firm from one country acquires another firm in another country. The importance of such an activity is that the construction industry becomes an increasing player in a nation's balance of trade. Essentially, contracts won abroad will bring in export earnings to a country, yet the importation of foreign building materials and services will result in a leakage from the domestic economy.

The construction industry tends to be highly labour-intensive and thus a major source of employment in the economy. Many aspects of construction, such as bricklaying and plumbing for example, would be difficult or impossible to mechanize, or at least it would not be cost-effective to do so in most circumstances, and therefore this high dependence upon labour is likely to continue. This is not the case in the building materials supply industry as cement works and brick works, for example, can be readily automated. Similarly, the advent of laser measuring equipment in the field of surveying can potentially reduce the demand for trained surveyors as site and building measurement can be done more easily and quickly.

In a mixed economy the construction industry plays a crucial role in achieving the government's social objectives. The government requires construction to fulfil its public sector housing programmes, to build hospitals, educational establishments and provide public buildings, such as libraries for example. However, in many countries there has been a noticeable reduction in the public sector's involvement in the provision of goods and services, and therefore the volume of work from the state sector for the construction industry has been declining.

Another feature of the construction industry is that apart from a small degree of prefabrication the product is generally not transportable. This is a feature that is perhaps unique to construction as the majority of other goods can be manufactured in one location and sold in another, and therefore any market disequilibrium can be rapidly resolved. This aspect of fixed location in construction means that shortages of certain types of building can exist in one area, yet an excess of floor space can exist in another.

The structure of the private sector of the construction industry

The following points are again generalizations and are not country specific. However, they broadly represent a summary of commonly observed features of the private sector of the construction industry.

- Predominantly small firms dominate the private sector of most construction industries.
- Although building demand in total is geographically dispersed, small, local contracts are the most common source of work for the construction firm in the private sector.
- The average number of employees in private sector firms is very low with many employing less than five people. This feature not only reflects the size of the firm, but also demonstrates the increasing tendency for the employment of subcontractors rather than direct labour especially in the case of large contracts.
- Most private firms tend to specialize in a certain area of construction such as plumbing, wiring or roofing, for example. Generalized data reveals that fewer than half of the firms observed classified themselves as 'general builders'.
- The 'failure rate' of small construction firms is very high. Such a high number of bankruptcies can be explained by a variety of factors but normally demonstrates the close relationship between prosperity in the construction industry and the performance of the economy at large.
- Despite the dominance, in terms of numbers, of the small firm, it is usual to find that a small number of very large firms are important in terms of overall construction industry output. In financial terms a major project undertaken by a large firm can be worth more than the total of many of the smaller projects undertaken by smaller firms. Moreover, the productivity of such large firms is far higher than their small counterparts.

This large number of small firms can be primarily explained by the fact that many of the forces that make it advantageous for a firm in manufacturing to develop into a larger-scale enterprise, are not present in the construction industry. The difficulty in the construction industry is that there are few cost advantages of expansion due to the lack of readily accessible economies of scale (see Chapter 8). Significant diseconomies of scale are unlikely because of the following factors:

- Production is based on individual sites rather than in a central manufacturing location. Thus, building firms cannot take advantage of the mass production techniques of most industries.
- There is a lack of standardization of the product. Each building is different reflecting differing requirements of clients, different locations and varying site conditions.
- There is a low use of large-scale capital as the conditions of construction are not generally favourable to the use and design of such equipment.
- The system of interim project finance means that large sums of initial financial capital are not always required to undertake the development process. Therefore, even small firms are in a situation to raise the start-up finance required for the initiation of most small-scale developments.
- It is unlikely that construction firms will achieve monopoly status by gaining control of materials suppliers, as building materials suppliers are themselves in a very strong position.

International construction activity

Some domestic construction firms may feel that they are sufficiently experienced to tender for projects abroad as well as in their local market. Foreign contracts will become attractive if the level of productivity from such work seems to be greater than that offered by building at home. However, it should be noted that there could be many problems with overseas investment especially because such an involvement often carries far greater levels of risk (see below). Despite this there has been a notable growth in the number of construction firms operating on a global basis. This growth in international work can be attributed to a number of factors as seen in the following points.

- During times of a depressed domestic market, overseas opportunities can look comparatively more rewarding assuming that there is not a complete global economic slump. Different parts of the world can experience economic prosperity and recession at different times. Indeed both the pattern and severity of a country's business cycle may well be partially dependent upon its stage of economic development. As a consequence, areas of the world that show great market potential now may not always do so. Conversely, there can be some promising markets in the future that are not yet established as areas exhibiting high levels of construction activity and growth.
- Due to the tremendous wealth generated in the oil-producing countries of the Middle East, the overall level of construction orders have, in the past, reflected the state of the oil market. Thus, when the price of oil is high, oil-producing nations will receive more income, which can then be spent on their development.
- The strong emergence of newly industrialized countries (NICs), especially in the Far East, has led to the very rapid development of these areas. Obviously countries that grow quickly require substantial investment in all forms of property so as to increase their necessary building stock.

In obtaining overseas contracts the construction firm must be able to demonstrate that it is in a better situation to undertake a project than a domestic firm in the target country. In relation to this point a number of key issues need to be raised, details of which now follow.

- The size of firm is normally a barrier to entry to overseas work. Research has shown that most work put out to international tender is concerned with very large projects for which only the largest firms are best suited to compete. Typically the problems of both project size and risk are partially reduced via joint ventures with other firms. Moreover, more experience and knowledge of a market can be gained by formally merging or co-operating with domestic firms, or by creating a local subsidiary.
- Foreign firms are normally in a better position to obtain contracts when there is a lack of indigenous construction technology and management skills in the country in question. Although small to medium sized projects can be undertaken by local firms they may lack the necessary expertise required for larger projects.
- The 'rules' of fair international competition are sometimes artificially distorted as governments lend assistance to their own nation's construction companies. In an effort to increase their chances of securing overseas orders governments can provide them with a number of assisted advantages, such as:
  - Providing state subsidies to the construction firm in order to reduce its costs.
  - Acting as guarantor so as to offset the potentially higher level of risk associated with much overseas work.
  - Introducing pre-conditions to any financial aid given to the recipient country so that it has to spend the aid that it receives on using firms and purchasing products from the donor country.
  - Putting diplomatic pressure on some countries in order to ensure that they award a contract to a specific country.
- To ensure enhanced profitability from international operations many multinational corporations (MNCs) have undertaken the practice of transfer pricing. Transfer pricing is effectively an accountancy procedure designed to lower the total tax burden of a multinational firm. Essentially it enables intra-corporate sales and purchases to be artificially invoiced so that profits accrue to those branch offices located in low tax countries, while subsidiaries in highly taxed countries show little, or no, taxable profit.

However, before becoming involved in overseas work evidence shows that construction firms must be made aware that a number of serious difficulties often occur with such contracts. These problems are perhaps most marked, and frequently encountered in the least developed countries (LDCs) of the world. In fact the existence of these problems in such locations increases the attractiveness of more stable markets in the more developed countries. It should be recognized that:

- International competition is very fierce, with firms from some countries benefiting from cheaper sources of project finance, lower labour costs and higher levels of construction productivity.
- Some large construction firms have exposed themselves to serious financial problems after going ahead with projects when they have received insufficient and unreliable local information. For example, it is important to get a clear picture on all aspects of the building process ranging from likely increases in building costs during the project's life to the eventual state of the market when the development is completed.
- Political turbulence in some countries has meant the total abandonment or delay of some construction projects. Moreover, in the case of a change in government, especially via military means, there have been examples of the new rulers refusing, or failing, to meet the commitments of the previous regime. As a result previously ordered construction projects may not be paid for on previously agreed terms, if at all.
- Financial crises have occurred whereby the overseas client has gone bankrupt and therefore cannot finance the ongoing development, or purchase it, once it is complete. Alternatively, if the building firm is working in conjunction with a group of local firms, financial difficulties with any one of these partners can jeopardize the whole consortium.
- High rates of inflation in many countries can seriously erode the value of projects agreed in local currency terms and can lead to large increases in building costs. The majority of projects take a number of years to complete, increasing their exposure to changes in the value of money.
- Foreign currencies can also depreciate in relation to other currencies reflecting the poor state of the local economy. Currencies are often officially devalued in an attempt to make the country's exports cheaper and imports more expensive (see the discussion on the balance of trade in Chapter 11). For example, one may negotiate the initial contract at an exchange rate of £1=$1, where the '£' symbol represents the currency of the home country of the construction firm, and the '$' symbol represents the currency of the overseas country in which the project is being undertaken. Therefore if the contract is worth $100m the firm should receive £100m when it repatriates its earnings to its home country (obviously these gross figures will be affected by taxation in both countries). However, if the $ is devalued so that the new post devaluation exchange rate is £1=$2, on conversion of profits at the completion of the development the construction firm would only receive £50m for its $100m.
- Rigid exchange controls can hinder the free flow of both project finance and the eventual repatriation of profits. Such controls are frequently highly bureaucratic in nature and can lead to costly project delays, or even the cancellation of potential projects. For example, an application may need to be submitted to a country's central bank so that money can be released to purchase essential building materials or services from abroad. If there is any delay in this procedure it could hold up the whole project because of the sequential nature of construction tasks. Otherwise, alternative second-best solutions may have to be tried.

- Some countries often seek to impose stringent conditions on foreign firms who gain access to their markets. For example, they may insist that a percentage of local suppliers and contractors are used, as well as promoting joint ventures with domestic firms. This condition should not necessarily represent a problem in itself, but construction firms have been let down by under-qualified local contractors or they have faced constraints in the local supply process such as an inadequate delivery of materials of satisfactory and consistent quality.
- National or local laws, customs and working practices can lead to misunderstandings and project delays.

Despite these potential difficulties it should be noted that they are variable over time and between countries. Moreover, the growth in the globalization of construction work suggests that these problems are not insurmountable and that the benefits frequently outweigh the negative aspects of such work.

The advantages to countries of awarding contracts to overseas construction companies are that these firms can provide both foreign skills and capital. This is of particular importance where indigenous firms may not yet be big enough, or sufficiently experienced, to take on all contracts, especially the larger, more complicated ones. However, there are also many criticisms levelled against multinational firms and their operations. For example, MNCs are unlikely to re-invest profits in the host country as they export them to the parent company in their registered country of origin. Moreover, elaborate structures built by such firms often necessitate the importation of materials and expatriate skills that can use up a country's valuable foreign exchange. In addition to this, research has showed that many multinationals do not greatly contribute to tax revenues due to their practice of transfer pricing. Indeed, the firms often benefit from the receipt of attractive tax concessions and subsidies that are offered to encourage them to undertake projects in the first instance. Finally, it is logical to fear that the success of foreign firms participating in domestic markets could inhibit the growth prospects of indigenous firms.

This chapter has been designed to give an insight into the variety of economic forces that shape the behaviour of firms in the built environment. The analysis has ranged from an examination of the costs and revenues that face individual firms, to the market in which they find themselves operating in. The application of the theory presented here enables one to judge how firms are likely to react under a number of circumstances, and how the firms can actually influence the market.

Part 3
The economy and its management

10 A simple model of the economy for the property analyst

The analysis so far in this book can be categorized as micro-economics because it involves the examination of individual aspects of the economy. For example, micro-economics enables one to examine the likely prospect of a small building firm entering a highly competitive industry (see Chapter 9), or alternatively, one could look at the market for a specific type of property such as housing, or office accommodation in a particular town (see Chapter 3). Now that the reader has gained this knowledge, and understood the methods involved, these issues can be brought together so as to examine the economy in its entirety. Therefore the work now moves on to study macro-economics.

This part of the book examines a variety of large-scale issues and their impact upon property markets and the construction industry, as well as the built environment in general. This chapter will examine the workings of the economy, whilst highlighting economic problems that can and do occur. The analysis will be kept to a relatively general level as it is felt that the application of the theories, rather than an intricate knowledge of their detail, is far more useful to the student of real estate issues who is unlikely to want to become an economic theoretician. As a logical extension, Chapter 11 suggests likely scenarios for firms in the property arena if the economy experiences difficulties such as high rates of inflation or unemployment. Finally, Chapter 12 investigates how governments attempt to manipulate the economy, at both the national and local level, in order to achieve economic and political objectives.

A macro-economic model

There are a wide variety of models available to explain how the economy operates. These models range from the very simple to the highly complex. However, as already stated, the objective of this book is to keep such models as basic as possible so that they can be easily understood and readily applied. Moreover, so as to utilize the knowledge gained so far, theories learnt in the micro-economics section of the book are now to be built upon to examine macro-economic issues and their relationship with the built environment. By adopting this more simplistic stance it must be appreciated that some loss of detail and accuracy may occur, although the basic principles expounded by the following analysis are sound.

As an initial starting point it must be emphasized that in an open and mixed (macro) economy, as is the situation in most countries of the world, there are four key components of the economy: consumers, firms, government and foreign trade. An open economy is one which trades with other countries and a mixed economy is where both the public and private sectors are involved in the actual workings of the economy. This chapter will now briefly discuss how these components of the economy interact so as to provide a simple working model of the economy itself. This model can be formulated from straightforward demand and supply analysis.

Demand in the economy

Consumer demand

Every day consumers in general will enter into a number of transactions ranging from the buying of necessities to the purchase of expensive consumer durables. Such demand will depend upon a large number of variables such as disposable income, government policy and the desire to save, to name just a few.

With respect to incomes, if consumers have a high propensity to consume it implies that they spend the large majority of their incomes and save only a little. In fact, as shall be seen shortly, with the introduction of the multiplier process, a useful economic concept is the marginal propensity to consume (mpc). The marginal propensity to consume is simply a measure of how much money consumers spend out of any increment in their incomes. For example, an mpc of 0.8 suggests that for every 1 per cent increase in income, consumers will spend 80 per cent of each 1 per cent rise, and thus only save 20 per cent of each 1 per cent rise. Therefore, as one can either save or consume, the marginal propensity to save can be found by subtracting the marginal propensity to consume from one.

> If: mpc = 0.8
>
> Then: mps = 1 − 0.8 = 0.2

It should be noted that a consumer's income could come from a variety of sources and not just from wages or salary. Income can also be gained from social security payments to the unemployed (where available), interest received from monies held in deposit accounts, dividends from shares, and rent from let property.

Looking at the influence of expectation-led behaviour, it is likely that consumers will delay consumption expenditure if they expect prices to fall in the future. Conversely they are likely to increase consumption if they expect prices to rise, or fear that shortages may occur. Unfortunately both of these forms of expectation-led behaviour can be self-fulfilling prophesies.

For example, if people increase their spending due to a fear of future shortages and rising prices, demand will go up thus putting upwards pressure on prices and existing supplies. In this way consumers are likely to cause shortages which may not have occurred in the first instance.

Chapter 12 deals in detail with the influence of government policy on overall consumer demand. At this stage, however, it is sufficient to state that government can attempt to influence consumer spending by altering the availability of credit, changing the cost of borrowing via adjustments in the interest rate, or by changing the level of taxes, for example.

Relating consumption expenditure to the built environment, it can be seen that the higher the level of consumer demand for goods and services, the higher the need for retailing and industrial output. Thus, increased demand will lead to a greater need for buildings (shops and factories) in which such economic activity is carried out. A problem arises in that the existing domestic supply of these buildings may not be able to cope with elevated levels, or sudden increases, in demand. This resultant supply constraint will come about if there is insufficient capacity in the current buildings to meet the new, higher levels of demand, as it will take time for extensions to buildings to be built, or new premises to be found. Moreover, new businesses may well need to be set up to accommodate this demand, and such new enterprises will not materialize overnight.

Therefore, in the immediate period one could expect, for example, overcrowded shops before more retail outlets were built. Furthermore, as it is likely that there will be little increase in short-run supply from the domestic industrial sector, this could lead to an increased volume of goods from abroad, imported by retailers in an attempt to fill the gap created by excess demand. These imports could worsen the country's balance of trade, yet the alternative is higher inflation as prices rise in an attempt to ration out current supply amongst the high levels of demand.

Another consideration, as shall be seen in the forthcoming discussion on the multiplier process, is that a high level of consumption expenditure is likely to lead to rising incomes (see later). As incomes increase, more people are likely to be able to enter the housing market, afford better, bigger houses, or have extensions built on to their existing houses. Therefore, one could expect increased activity in the residential sector as well as that seen in the industrial and retailing sectors.

In other words, increasing levels of demand tend to encourage activity in property markets, perhaps leading to a property boom. However, it must be noted that the overall impact of rising demand on the property market can be severely distorted by social factors. First, it has been observed in some countries in the past, that increases in incomes have largely been spent on housing rather than any other sector of the economy. This primarily reflects a desire for owner occupation, and means that manufacturing industry, apart from those industries directly involved with the housing sector, such as furniture and carpeting for example, do not experience any great changes in demand and corresponding growth. In such circumstances one would expect there to be a boom in both the residential market and some parts of the retailing sector of the property market, but not in the industrial sector in general.

A second example that could lead to little growth in the industrial sector would be that some countries have a very high propensity to import. In such instances people have a preference for imported goods over those produced domestically. The reasons for such preferences can be varied and range from the fact that it is simply fashionable to own imported goods, or that some imported goods are deemed to be better in terms of quality than similar locally manufactured goods. If there are high levels of import penetration, increases in incomes are likely to be spent on foreign goods. Therefore, heightened levels of domestic consumption expenditure will tend to benefit the industrial sectors of the country's foreign trading partners rather than the industrial property market of the home country.

Finally, it should be noted that consumption expenditure patterns and levels can be influenced by long-run structural changes in the economy such as changes in the distribution of income, or demography. For example, if there were a significant level of internal migration within a country from a depressed region to a more prosperous one, there would be growth in the property market, in terms of prices and new building, in the region to which people were migrating. Conversely there would be a relative collapse of markets in the area experiencing a net outflow of people. This would have an impact upon all types of property as not only would more housing be needed in the preferred region, but more people moving to that area would also be likely to stimulate retail and industrial growth in the area due to the concentration of their purchasing power. The selling price, or rent levels, of existing buildings in this area would increase reflecting the increased competition for their occupation. However, the depressed region would become even more of an economic backwater. As people moved away, the demand for houses would drop and house prices would decline. Moreover, a great deal of the area's spending power would be lost to the prosperous region, and as such retailers and industrialists in the depressed region may need to reduce output, or even close down.

Demand from firms

Obviously firms themselves need to demand goods and services from other firms in order to function as a business. Thus, a large number of transactions in an economy occur within this sub-sector. For example, in order to construct buildings, a construction firm will order raw materials from material suppliers, and may also need to purchase new machinery as existing equipment wears out and needs to be replaced. If demand in the economy grows during a period of economic recovery, one would expect more building work to be ordered and thus construction firms, and the businesses supplying these firms, should be faced with rising orders. Note that in both of these instances the demand is a derived one, i.e. firms do not order these items as an end product, but require them to produce the buildings for which there is a final demand. This is again a process that will be stimulated by the multiplier effect (see later). However, it should be noted that much of any inter-firm demand may leak abroad as construction firms, for example, order materials and components from overseas suppliers. Such a propensity to import by the industry may be influenced by quicker delivery times, better quality, improved choice, etc.

Demand from government

A large volume of overall demand in most countries comes from the public sector at both the local and national level. The volume of such demand will depend upon a variety of factors, with the two main forces being:

- The current state of the economy and the government's reaction to it. For example, a country experiencing a deep economic depression may need strong government intervention to put it 'on the road to recovery' (see Chapter 12).
- The political philosophy of the government in question. Some governments are very keen to promote largely unrestricted economic activity by the private sector, whereas others believe in a greater degree of public involvement.

In most countries, however, the state sector will be involved in the provision of the main road networks, constructing educational establishments such as schools and universities, and building prisons, hospitals, defence installations, etc. Therefore, construction firms, in particular, will receive a large volume of demand from this sector. Again a multiplier effect will be apparent (see later).

Demand from abroad

Another potential source of demand in the economy comes from abroad. That is, other countries will demand goods and services from the home economy. In order to attract, or maintain, such demand, a nation needs to ensure that its goods and services are internationally competitive in terms of both price and quality. This could be achieved by competing on the grounds of efficiency and technological lead for example.

Good management of the exchange rate by government is also required so as to ensure that the country's exports are competitive in terms of price. However, it should also be noted that although a country may win many export orders, much of the demand created by such activity, and its associated positive impact upon the economy, may be reduced or even eliminated by domestic demand leaking abroad to buy foreign goods. Therefore, the net effect on the country's economy will depend upon the country's overall balance of trade.

In terms of construction, export orders are won by domestic firms securing contracts overseas. However, the net effect on the economy of such activity by the construction sector would have to be viewed against all the materials and services that the building industry imports from abroad.

A conclusion on demand in the economy

In order to measure the overall level of demand in an economy, all the sub-sectors discussed above need to be examined. Once this is achieved one can attempt to total them together in order to create an aggregate demand function for the whole economy. Aggregate demand can be expressed in notation form as:

$$AD = C + I + G + (X - M)$$

Where: AD = aggregate demand
　　　　 C = consumption demand
　　　　 I = investment demand (demand between firms)
　　　　 G = government demand
　　　　 X = exports
　　　　 M = imports

Supply in the economy

As with all analysis, it is vitally important to consider both sides of the market. That is, meaningful analysis requires an examination of the supply side of the economy as well as the demand side. Therefore, the text now progresses to examine the sectors of the economy that supply goods and services. Once this is achieved an aggregate supply schedule can be created using exactly the same methodology as with the earlier aggregate demand analysis.

Supply from consumers

Individuals effectively supply on to the market either their labour or their entrepreneurial skills. Therefore, the amount of people willing to work in an economy at a given set of wages, or reward, will have a great influence upon the overall level of supply in that country. For example, in an economy affected by lethargy, a poor state of general health, or compulsory military service, a large proportion of the labour force will not be able to provide their productive services. As a consequence overall economic output will be impaired by a reduction in the supply of labour. Likewise, a country that does not invest in an adequate education system may fail to produce entrepreneurs to initiate, or advance, the supply process. In fact, it may lose entrepreneurs if it fails to give them sufficient financial or material incentives to remain in their home country. This problem can sometimes be seen as a 'brain drain' as top professionals leave their country to take up favourable circumstances in another country, such as lower levels of personal taxation, or the provision of superior research facilities. Without a good education system, the country may not create a sufficient pool of skilled labour that would be required to undertake many forms of production.

Supply from firms

Supply from firms is perhaps the most obvious form of supply as it is the firms that provide the goods and services that we demand as consumers. The success, and therefore the output, of such firms will depend upon a variety of key economic variables including the commercial taxation structure, the state of the economy itself, the cost of finance, and the structure of the industries themselves. For example levels of output will partially depend upon whether firms operate in a highly competitive market structure or a more monopolistic one (see Chapter 9).

Government supply

As shown in the section on aggregate demand (above), the government is frequently involved in the provision of a large number of goods and services ranging from the operation of nationalized industries, to the provision of public utilities.

Imports

Another source of supply into the economy is the volume of imports from abroad. If a country has good export performance, it can afford to make up any inadequacies in domestic supply via this route. However, if imports constantly outweigh exports the economy may soon suffer from a balance of trade crisis. In the construction sector one can witness imports in the form of imported materials and equipment. In addition to this foreign firms can win orders in the domestic economy.

A conclusion on the supply side of the economy

Once output has been measured from all of the above sectors they can be summed together to generate an aggregate supply function.

A simple demand and supply model of the macro-economy

Given full information concerning both aggregate demand and aggregate supply a simple demand and supply model of the macro-economy can be created that will:

- Illustrate a framework with which to understand the workings of the economy as a whole.
- Provide a means specifically to analyse the impact of government intervention upon the economy.
- Demonstrate the impact of external influences on the domestic economy.

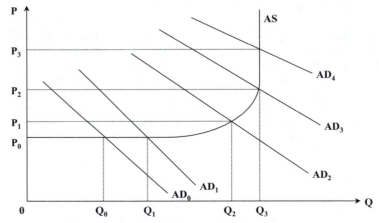

Figure 10.1   *A simple model of the macro-economy*

This model is represented in Figure 10.1. In this diagram the curve AS represents the aggregate supply function, and the various curves labelled AD represent differing levels of aggregate demand that may be experienced, ranging from the low demand depicted by the curve AD_0 to the very high level of demand shown by the curve AD_4. Here, it can be seen that, given the existing aggregate supply function, changes in aggregate demand will cause changes in the level of output in the economy (Q) and in the general level of prices (P). With the economy represented in this manner the following points should be borne in mind.

1 It can be seen that the aggregate demand function is downwards sloping. This reflects the normal inverse relationship between price and quantity as people economize or postpone consumption if prices are too high (see Chapter 1).
2 Aggregate demand can shift depending upon the variety of influences that affect consumers, investors, government decision-makers and exports as suggested above. For example, if aggregate demand were depressed during an economic recession, it would be represented by a low aggregate demand function such as AD_0. However, as the economy recovers the curve would shift to the right representing higher levels of overall demand.
3 Although a nation's ability to produce, and therefore the position of its aggregate supply curve, is relatively fixed in the short run, it can and does change over time. For example, increases in aggregate supply would cause the aggregate supply function to shift to the right. Such movement could be due to effective government supply-side policies (see Chapter 12) or technological change. Conversely environmental concerns about current methods of production could prevent the techniques being used which could retard the aggregate supply function to the left.

4 It can be seen that three distinct sections characterize the aggregate supply function. These sections range from an initial perfectly elastic portion at low levels of output, to a perfectly inelastic portion at higher levels of output.

5 The shape of the aggregate supply function suggests that at very low levels of output, any increases in demand can be accommodated by utilizing excess capacity and therefore there will be no inflationary pressure. This can be seen where aggregate demand increases from AD_0 to AD_1. This shift in demand produces a rise in national output from Q_0 to Q_1, but prices remain unchanged at P_0. In terms of the built environment, such a scenario might occur during a recession or period of low demand. In these circumstances it is highly likely that there will initially be vacant and under-utilized industrial and commercial properties. Therefore, as aggregate demand increases, and more industrial and commercial property is required, firms may be able to utilize existing accommodation if it is of an appropriate form. In other words they are unlikely to be vying for a limited amount of floor space as such competition would put upwards pressure on both the sale price and rents of properties.

It should be noted, however, that this broad discussion relates to the economy in general and therefore may disguise the fact that some regions of a country could suffer from a lack of usable buildings. Alternatively, other relatively prosperous areas could be better suited to weather a recession and its correspondingly depressed demand. With both of these scenarios any growth in aggregate demand will tend to lead to competition for space. This will then produce a resultant inflationary impact as buildings are sold or rented to the highest bidder. However, these cases are either unusual, or they do not significantly deviate from the norm. As such they do not greatly influence the overall general picture of the economy. In fact if a region's performance differed to a very large extent from that of the rest of the economy, it could be argued that it should be excluded from any global analysis, and treated as a special case. Otherwise its inclusion could distort the potentially representative average.

If demand were to rise further, say from AD_1 to AD_2, some shortages of supply could begin to occur, as the aggregate supply function becomes relatively inelastic. Such inelasticity is a result of suppliers having difficulty in quickly responding to increases in the level of demand. Once the economy has reached this stage the available supply will be rationed out by the price mechanism leading to an increase in the general level of prices from P_0 to P_1. In fact, if aggregate demand were to rise too rapidly, or too far, such as a change in the level of aggregate demand from AD_3 to AD_4 for example, it would produce a scenario whereby output would remain constant at Q_3, yet prices would rise from P_2 to P_3. Such inflationary pressures could be reduced if action was taken to boost aggregate supply by building more factories, increasing productivity, etc. This policy, if success-ful, would shift the aggregate supply function outwards to the right (see Chapter 12).

Therefore, neither extreme of low, or high, demand is good for the nation. At very low levels of demand, such as AD_0, the economy would experience little, or no, inflation but would suffer from the opportunity cost of a high under-utilization, and thus underemployment, or resources. In such a situation one would expect to see evidence of an economic depression such as derelict factories and vacant office buildings, as well as people being out of work. Alternatively, at very high levels of demand, such as AD_3 and AD_4, factors of the production, such as capital and labour, could be fully employed given the current state of technology, and accepted procedures. However, the cost would be high inflation such as that shown by the price levels at P_2 and P_3.

As changes in aggregate demand and aggregate supply can have such far-reaching consequences on the economy, most governments seek to exert some form of control on these two forces. To enable governments to attempt such intervention they have at their disposal both demand management policies and supply-side economics. Before examining such government manipulation of the economy, and its implications to the property and construction industries, a good understanding of the multiplier process is required.

The multiplier process

The multiplier, as the term suggests, is a process that magnifies (multiplies) the net impact of any alteration to the economy. The multiplier effect can work in both an upward, positive manner, as well as in a downward, negative manner. For example, if government were to increase an injection into the economy, such as a rise in government expenditure, it would be found that the multiplier process would be positive and increase incomes by a greater amount than the initial value of the injection. Conversely, if there were to be an increase in a leakage or outflow of the economy, such as increased expenditure on importing foreign goods, the multiplier would be negative and would tend to create a downward spiralling effect of declining incomes in the economy. To understand the mechanisms involved in this process, examples for both the positive and negative multiplier appear below.

The positive multiplier

Positive multipliers are induced when there is an initial injection of money into the economy. Such injections can be created by an increase in government expenditure (G), an increase in the level of investment (I), or an increase in the volume of export orders (X). For example, if the government were to increase its expenditure on the repair and maintenance of public buildings, the mechanisms behind the multiplier process would suggest that the overall impact on the economy would be far greater than the initial rise in government expenditure.

The reasoning behind this result can be demonstrated by going through the following sequence of events: if the government spends more money on public buildings, such as schools and libraries for example, construction firms would have to be employed to carry out the work. These construction firms would then have to order materials from firms in the material supply industry. Subsequently firms in the materials supply industry may need to take on more workers to cope with the increased orders, and the building firms themselves may need to employ more labour so that they can undertake the new contracts. As a consequence more people are employed in both the construction industry and the materials supply industry. Some of those who now have a job in one of these industries may have been previously unemployed, so that now they are earning an income whereas before they were not.

Thus, with greater numbers in employment increased spending in the shops would be expected as more goods and services are purchased. The retail sector will then need to order further supplies from industry as stocks are depleted, and they may also need to take on more staff to accommodate the increased sales volume. These new staff employed in the retail sector will spend their money in the shops, and another similar sequence of events is initiated. Thus, it can be seen that consumer prosperity can directly lead to enhanced activity in both the retail property market and the industrial property market. Indeed, as more are employed and total incomes increase, there is also likely to be an increase in activity in the residential property market as more people feel that they can enter the owner-occupied market or move up the housing ladder. Therefore, in conclusion, any injection of money into the economy is highly likely to lead to a far greater level of net income due to the series of chain reactions that such an injection will set off.

It can now be seen that the level of consumer expenditure is central to the multiplier process. Therefore, when quantifying the magnitude of the multiplier, analysts need to find information about consumption behaviour as well as knowing the actual monetary value of the initial injection. Of particular use is an understanding of the consumer's marginal propensity to consume (mpc). That value illustrates how much consumers will spend upon consumption due to any rise in income. Once such data has been obtained, the following simple formula can be used to ascertain the actual value of the multiplier:

$$K_u = \frac{\Delta \text{ injection}}{(1 - \text{mpc})}$$

Where: K = the symbol for the multiplier
u = the notation to indicate that the multiplier is positive and creates an upwards movement in the economy
Δ = denotes a change in the level of the injection

For example, assuming an injection of £100 million of government expenditure, and that the marginal propensity to consume in the economy is equal to 0.6, one could simply insert these values into the equation in order to ascertain the likely overall magnitude of the multiplier process:

$$Ku = \frac{\Delta G}{(1 - mpc)} = \frac{£100}{(1 - 0.6)} = + £250m$$

Where: G = government expenditure

The result of the formula shows that although government expenditure was only increased by £100 million in the first instance, a sequence of events was initiated causing the overall level of activity in the economy to grow by a further £150 million to a total of £250 million. Thus, as the value of the initial injection has been multiplied by two and a half times, in this example, the value of the multiplier is said to be 2.5. Alternatively, one could find the value of the positive multiplier in the following way and then multiply the answer by the level of the injection:

$$Ku = \frac{1}{(1 - mpc)} = + 2.5$$

Knowledge of the value of the multiplier is important for the government in terms of promoting growth and employment, but also when trying to control inflationary pressures in the economy. For those involved in the property world, an idea of the magnitude of the multiplier is useful when trying to determine the level of the new development that may be required from any stimulation in activity caused by increases in injections into the economy. However, it must be appreciated that the full impact of any positive multipliers may be counteracted by leakages from the economy creating the potential for negative multipliers to exist.

The negative multiplier

Negative multipliers are induced when there is a withdrawal, or leakage, from the domestic economy, which effectively reduces the level of income in that economy. Examples of such leakages are listed below.

● Increasing levels of taxation (T) reduces people's spending power. As money is taken away from consumers in the form of taxes such as income tax or even local property taxes, they will have less disposable income with which to spend upon goods and services. As a consequence consumer demand is likely to fall.
● Expenditure on imported goods and services (M) means that consumer demand does not benefit local businesses. Therefore, the positive impact of such expenditure is felt abroad rather than at home.
● If people were to save (S) a higher proportion of their income than they do at present less money would be available for expenditure and therefore demand would fall. The degree of savings can be measured by the marginal propensity to save (mps). This measure shows how much additional income is saved rather than used for consumption expenditure. However, in the long run, increased rates of saving may well lead to higher levels of consumer demand. The logic behind this statement is that

if savings are wisely invested they will create a return that can enhance the spending power of the saver in the future. Moreover, deposits in financial institutions in the form of savings allow such institutions to lend that money out to others such as firms wishing to invest.

The mechanism behind the negative multiplier is very similar to that of the positive multiplier, except that it occurs in the reverse direction. This can be seen in the following way: as a leakage increases and consumer demand falls, less will be spent by consumers on goods and services. Therefore, for example, the retail sector may record a drop in sales, and offices will conduct less business. If this does occur these businesses are likely to lay off workers who are no longer required, and order less from their suppliers in the manufacturing sector. As the demand for industrial output declines, firms in this sector may also need to make some staff redundant. As more people become unemployed, they too spend less in the shops, and as such the downward spiral continues.

For example, if taxes are raised, perhaps to finance increased government expenditure, consumer's disposable income will be reduced and thus so will their spending power. Therefore, a negative multiplier will be induced. In an attempt to qualify the net downward effect caused by this mechanism, one again needs to gather information on the value of the current marginal propensity to consume, and the actual level of the initial withdrawal (which is a tax rise in this example). Once this data is obtained it can be put into the following formula which would show the overall magnitude and net effect of such a leakage:

$$Kd = \frac{- \text{mpc} \times \Delta \text{ leakage}}{(1 - \text{mpc})}$$

Therefore, for example, if the overall tax burden on the economy had risen by £80 million, and the marginal propensity to consume was again equal to 0.6, these values could be inserted into the following formula in order to estimate how much of a deflationary effect this would have on the economy:

$$Kd = \frac{- \text{mpc} \times \Delta \text{ T}}{(1 - \text{mpc})}$$

$$Kd = \frac{- 0.6 \times £80m}{(1 - 0.6)} = £120m$$

Where: T = taxation

Therefore, because of the sequence of events that are set off by the negative multiplier the leakage has produced pressures in the economy that will lead to a decrease in overall income. The formula indicates that this decrease will be in the order of £120 million rather than just the £80 million of the initial increase in the tax burden. In other words, the net effect is 1.5 times greater than the actual leakage. Thus, in this example, the value of the negative multiplier is 1.5. Alternatively, one could have found the value of the negative multiplier in the following way and then multiplied this answer by the level of the initial withdrawal:

$$Kd = \frac{-mpc}{(1 - mpc)}$$

Thus, in our numerical example from above:

$$Kd = \frac{-0.6}{(1 - 0.6)} = -1.5$$

Likewise, if £20 million has been spent on imports, the formula would suggest that because of the sequence of events that are triggered off by such leakages, the overall effect on the economy would be far more damaging than this initial figure would indicate, as domestic industries may close down for example. Again one could calculate the full impact of such expenditures by inserting the relevant values into the formula:

$$Kd = \frac{-mpc \times \Delta M}{(1 - mpc)}$$

Where: M = imports

$$Kd = \frac{-0.6 \times £20m}{(1 - 0.6)} = -£30m$$

Therefore, direct leakages will create a negative impact on the overall level of income in an economy. Such a downturn in activity could also be caused by an increase in the level of savings, as any increase in the marginal propensity to save will cause a reduction in the marginal propensity to consume. In other words, less money will be available to consumers to spend upon present consumption. Thus, in our example, if the marginal propensity to save increases from 0.4 to 0.5, it would imply that people now saved 50 per cent of their income rather than 40 per cent. If this were to occur it would reduce the strength of the upward multiplier.

As seen in the initial example, if government expenditure was increased by £100 million and the marginal propensity to consume was 0.6, the net effect on the economy would be an increase in incomes by £250 million, or two and a half times the initial injection. However, if the marginal propensity to save were increased to 0.5, the marginal propensity to consume would drop to 0.5 so that the increase in government expenditure would only increase by a magnitude of two rather than two and a half:

$$\text{Original situation:}\quad Ku = \frac{£100m}{(1 - 0.6)} = + £250m$$

New situation: $Ku = \dfrac{£100m}{(1 - 0.5)} = + £200m$

Therefore, an increase in the marginal propensity to save means that less money is being spent in the shops, retailers are ordering less from industry, and so on. Note also that such a decline in the marginal propensity to consume caused by higher savings would also reduce the magnitude of the negative impact of the downward multipliers. This can be explained by the fact that less would be spent upon leakages such as imports, and less would be received in the form of sales tax.

A conclusion on the multiplier effect

It is highly unlikely that an injection into the economy, or a leakage from it, would occur in isolation. In reality, leakages from an economy are often offset by corresponding injections. For example high taxation gives a government more revenue in order to increase government spending; high levels of savings means that financial institutions have more money to lend out for the purposes of investment; and by receiving monies from exports a country is more able to pay for imports. Therefore, one needs to examine the full range of alterations in the economy so as to appreciate the net effect. For purposes of illustration, imagine that the government had increased its expenditure by £100 million, and in order to help finance this they have raised taxation by £80 million. Moreover, assume that £20 million of this rise in government expenditure is to be used for the purchase of foreign goods. It is assumed that the marginal propensity to consume has returned to its original value of 0.6.

The net impact on the economy could then be calculated by using the multiplier formula extended to cover changes in all variables:

$$\text{Net impact} = \frac{\Delta G}{(1 - mpc)} + \frac{-mpc \times \Delta T}{(1 - mpc)} + \frac{-mpc \times \Delta M}{(1 - mpc)}$$

$$\text{Net impact} = \frac{£100m}{(1 - 0.6)} + \frac{-0.6 \times £80m}{(1 - 0.6)} + \frac{-0.6 \times £20m}{(1 - 0.6)}$$

$$\text{Net impact} = +£250m - £120m - £30m = +£100m$$

Given a knowledge of these forces the professional in the property market should be able to gauge the likely impact upon the economy, and thus the property market, of any changes such as a rise in imports, an increase in taxes or a fall in government expenditure.

11 Primary economic objectives and the property sector

This chapter aims to examine the main economic objectives of government and how they may be achieved through macro-economic management. This analysis is set against the backdrop of property markets so as to see the impact upon the real-estate sector of an economy if such objectives are not met, or if any one objective has a particular emphasis in government policy.

In order to encourage a strong economic climate, and to maximize the well-being of the those who live and work in the economy, it is felt that a number of key macro-economic objectives need to be achieved. These objectives are normally:

- a steady rate of economic growth
- a low and stable rate of inflation
- a low level of unemployment
- a balance of trade.

Each of these goals are now briefly examined although the above is not an exhaustive list. A balance of payments and exchange rate policy are just two additional areas not covered in this text but worthy of further investigation. In addition it must be appreciated that government also has non-economic objectives which can act as a constraint upon pure economic objectives. For example, high levels of growth may be partially compromised to protect the environment.

A steady rate of economic growth

Economic growth implies that the level of activity in an economy increases over time as more transactions take place and more is produced. Such growth will cause an increase in the demand for goods and services in general, and more people are likely to be employed in an attempt to see that supply meets the new, higher levels of demand. As more people are employed, they receive an income, and therefore their expenditure is also likely to rise, and the multiplier process begins to operate (see Chapter 10).

This positive growth is beneficial to the construction industry, for example, in a number of ways.

1 Increased activity in the commercial sector is likely to create increased orders for office, retail and manufacturing buildings, although the levels of these orders would depend upon the degree of excess capacity that exists in the economy, and how much demand leaks abroad to foreign suppliers.
2 A healthier economy will give rise to higher incomes which would stimulate the residential new build market as more people can afford a home of their own, or move to a larger house. Thus, property surveyors are likely to benefit from increased levels of activity in the housing market.
3 Higher incomes enable the owners of buildings to spend more upon the repair and maintenance of their existing building stock.

If a steady state of economic growth is not achieved the economy will be characterized by varying levels of growth over time. Sometimes 'growth' could be negative as the economy goes into absolute decline. If either an economic decline or stagnation is observed, the construction industry will suffer a reduction in orders, as it will largely only be catering for the essential replacement of buildings as well as being involved in a basic level of repair and maintenance. Such a state of economic decline could have long-run effects on the built environment as property developers and entrepreneurs lose their confidence in the economy to achieve a favourable investment climate. Moreover, if an economy does experience negative rates of economic growth, many firms will either go out of business or will be operating below their full potential. In such circumstances one would observe an excess supply of buildings in the commercial sector and at the extreme dereliction would occur (see the discussion on the theory of the firm in Chapter 9).

It is important to appreciate that although economic growth is desirable it must be a steady and known rate that is not too slow, too fast, or too variable. If the level of growth were too slow the economy is unlikely to attract foreign investment as investors see the likelihood of more promising and higher returns elsewhere in the global market. On the other hand, if the rate of growth was very high it is likely to lead to inflationary pressures in the economy. These pressures could be detrimental to the long-run prosperity of that economy (see the discussion concerning the inelastic portion of the aggregate supply curve in Chapter 10 as well as the debate on inflation directly below). Moreover, fluctuating rates of growth are also likely to deter the investor even though the long-run trend may be encouraging. Investors need to be cautious in an environment where large sums of money are being risked, and the timing of returns is crucial. Fluctuating rates of growth lead to uncertainty about the exact position of the economy in the future with respect, for example, to the level of demand and inflation. Such uncertainty complicates investment in property, which by its nature is concerned with long-term returns.

It should also be understood that although a growth figure is reported for most countries from year to year, the statistic could be highly misleading. Any average figure can hide the fact that enormous variations in growth can occur within any economy. Certain geographical areas may be in decline, perhaps due to the closure of major traditional industries, whereas other regions may be experiencing great economic fortune. Thus, economic growth can lead to increasing inequalities in a country, as some benefit from it more than others and indeed some may not benefit at all.

Achieving a steady state of growth is far from easy and has proved to be illusive for many governments. Carefully formulated plans for growth have frequently been nullified by the emergence of exogenous shocks to the economy. For example, an unanticipated rise in world energy prices can unexpectedly dampen the prospects of any energy-dependent nation. Furthermore, policy-makers must ensure that neither the poor implementation of policy, or the failure of policy itself, actually retards the level of growth that would have occurred anyway in the absence of public intervention.

Despite striving for a steady state of growth most economies have been characterized by cyclical economic activity (see Chapter 12 and Figure 11.1). Observations of past macro-economic data have revealed that there are both short-run and long-run cycles. This pattern of economic behaviour is often termed 'the business cycle'.

Typically, the short-run cycle exhibits fluctuations in economic growth with an average life for each cycle of around four to five years. The length of each cycle is termed its wavelength. The amplitude of these cycles is normally quite small, i.e. the difference between the low point of the cycle and the high point is rarely that great. However, it must be noted that there are exceptions to this general pattern as economies have experienced major recessions and periods of exceptionally high economic activity. Significant disturbances are, however, relatively rare. Generally, the normal fluctuating short-term economic cycles can be caused by minor adjustments such as changes in the interest rate. Amendments to the cost of borrowing inevitably have a knock-on effect on property markets by creating additional pressures on those markets. For example, a rise in interest rates may lead to a reduction in the level of occupier demand as firms either attempt to reduce costs by making space savings or go out of business altogether. Such a decline in the demand for floor space will inevitably filter through to a reduction in the rate of development and a fall in both rental levels and property prices.

Short-run cycles seem to exhibit cyclical behaviour in themselves in that they appear to trace longer-term cycles with a duration of twenty or thirty years. Major structural changes or developments in an economy over time can cause these long-term waves. Demographic change is commonly cited as a key explanatory variable for long-term cycles. For example, changes in the birth rate caused by 'baby booms' naturally tend to have a re-occurring effect. A large increase in the numbers born one year will lead to another influx of new born some twenty years later as the original babies become adults and begin to have children of their own.

Another source of long-term change is the increasing rate of change in the field of technology. Technological change can take place in all parts of the economy. For example technological advances in the arenas of infrastructure and transport, or information technology (see Chapter 3) can lead to improvements in economic prosperity as well as movements in the geographic location of economic activity. In other words it is highly likely that regional differences will exist in terms of an area's relative position in the overall business cycle, i.e. some prosperous locations may hardly feel the impact of a recession. Conversely, areas of economic deprivation may not experience the positive impacts of a boom in the overall national economy.

Institutional change can be seen as another explanatory factor in the creation of longer-term cycles. Such changes take a long time to come to fruition and are therefore by definition only likely to have an impact in this broader cycle. For example, the level of economic activity in an economy has the potential to radically alter if the country opts to join a trading block or adopts a regional currency to replace its own domestic one.

An examination of the business cycle will also show that as economies tend to grow over time their business cycle oscillates around an upwards growth path. Therefore, by logical deduction, it must be borne in mind that some recent depressions will actually exhibit higher levels of economic activity than previous booms (see points A and B on Figure 11.1).

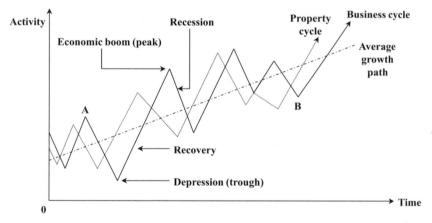

Figure 11.1 *The property cycle and the business cycle*

Observing cyclical patterns of similar wavelength and amplitude may encourage the belief that behaviour of the economic cycle is predetermined. However, it must be recognized that business cycles are influenced by both random and predictable events. Random events are normally exogenous and are, by their nature, uncontrollable from within the domestic economy. Predictable events are created by factors such as the legal and constitutional framework of a country, its political make-up or its ruling economic system. Disturbances can come from a variety of sources, some of which are briefly listed below:

- Consumer beliefs and their subsequent behaviour can change over time. For example, in many societies the borrowing of money (debt) used to be deemed as an embarrassment or even an evil. Nowadays, most accept short to medium term debt as part of their ongoing personal financial management.
- On the supply side of an economy, investors' expectations of the future can be self-fulfilling. For example, if investors are optimistic and believe in future prosperity they are likely to invest more. As a consequence, via the multiplier effect (see Chapter 10) such an injection of funds into an economy will cause that economy to improve. Conversely, investor pessimism can create the necessary climate for a decline in the level of economic activity.
- Financial deregulation and enhanced competition in the financial sector can release more forms of finance and at a lower cost. This in itself will lead to a greater take-up of borrowed funds which is likely to enhance economic growth via increases in both consumer and investment demand. Conversely, if banks have suffered reductions in profitability, say due to having to make provisions for bad debt, the amount of borrowed funds available on the market may be reduced. A reduction in credit would stifle economic growth.
- Technological change, as previously discussed (see above as well as the debate in Chapter 3) has the power fundamentally to alter the very foundations of an economy be it in the form of IT, industry or transport for example.
- The actions of international currency speculators can lead to alterations in the pattern of world currency exchange rates. If such changes are significant and occur over a short period of time they can destabilize whole economies and the general structure of international trade. Essentially the internationalization of investment and the beginnings of a true global economy are both events that foster the possibility of problems in one market being exported to another.
- International conflicts destabilize economies by creating a breakdown in trade between countries or can cause the closure of major trading routes. Economies that find themselves in armed conflict at home or abroad have to devote much of the nation's resources to financing such a campaign. Any diversion of funds into military expenditure could represent a significant opportunity cost for the rest of the economy.
- Sudden and significant changes in the price of essential raw materials and energy can have a major impact upon economies that are highly dependent on the import of such items. For example, an escalation of oil prices on the world market could radically increase the costs of energy for all in the economy. Oil prices may rise if oil-producing nations agree to reduce the supply (see the impact of supply shifts in Chapter 1). As consumers have to spend more on petrol, heat and lighting they will have less income at their disposal with which to buy other goods or services. In the same way, high-energy bills for firms could lead to a reduction in profitability and closure. Both of these results will create a reduced level of investment demand. If the changes are more gradual it is more likely that they can be absorbed by the economy as people have the opportunity to adjust.

- Sudden institutional change such as the collapse of government or a significant deviation in its economic policy could destabilize an economy. It is for this latter reason that many governments are reluctant to admit to a 'U-turn' in policy as this could affect investor confidence in their judgement and ability to effectively manage the economy.
- Changes in a country's national boundaries can also affect growth. A country can be spilt by the rise of independent autonomous areas. Alternatively it may increase in size due to the acquisition of other land through international agreement or as a result of conflict. The creation of an empire would be the extreme form of the latter example.
- Structural problems could prolong cycles, making the economy slow to react to policy. Poor infrastructure or a reliance upon outmoded traditional industries are examples in this instance.

Inevitably there is a strong relationship between business cycles and property cycles as most forms of economic activity occurs in buildings of one form or other. Indeed there is likely to be a two-way link between these cycles. For example, in terms of the economy influencing property, the demand for commercial buildings stems from the general level of aggregate demand and economic activity. In terms of the property market influencing the general economy there is the possibility that mass investment into development will initiate a multiplier effect that boosts the whole economy. Conversely, a decline in property market activity will tend to lead to falling property prices which, in turn, forces firms to reduce investment into the economy as their assets become worth less. Effectively a reduction in the worth of a company's assets represents a reduction in the collateral that can be offered to secure borrowed funds.

Property cycles and business cycles are unlikely to operate in complete harmony with one another. It is debatable as to which one leads the other. Although, mainly due to the sheer size of the property sector as a proportion of the economy as a whole, there is an obvious and logical correlation, the direction of causality is unsure. Indeed the direction of causality may not always be the same. Moreover, the clarity of the relationship is further complicated by the emergence of the internationalization of investment.

Property cycles are primarily influenced by:

- Alterations in influencing variables from outside the property market itself. Changes in government macro-economic policy are an example of such an exogenous force.
- Policy in related institutions such as those in the financial sector. The availability, choice and cost of finance are obvious examples.
- Factors inherent in the development process such as the time lag between the demand for a property and its completion.
- The beliefs and behaviour of those working within property such as valuers and developers. For example, over-pessimistic valuations of current property are likely to deter·future development and development finance.

The property cycle exhibits the following characteristics as it moves from one phase to the next. During a period of recovery there will be increases in the level of aggregate demand enhanced by the stimulation of the multiplier process. Correspondingly higher levels of economic activity will produce a derived demand for new buildings once relevant excess capacity has been absorbed. As a consequence, vacancy rates fall, and rents and capital values increase. These property indicators stimulate new development and thus a demand for more building land.

Property booms can cause pressure for interest rates to rise as an escalating demand for property can fuel inflation through higher property prices. Moreover, an increased reliance upon imported building materials, as domestic sources are exhausted, can adversely affect the balance of trade (see later). Indeed some commentators argue that high rates of activity in the property market can deter growth in the rest of the economy as investment for other uses such as equipment is diverted into buildings.

As already stated, increased demand leads to rising rents as firms begin to require more space within which to increase output. Although there is often much short-term enthusiasm and action by developers, the supply of new buildings should depend upon a longer-term view as the duration of the building process can straddle more than one cycle. A logical extension to this argument is the view that the current rate of interest is not necessarily a leading factor in the development decision. Rather, predicted variables are more relevant such as the future cost of borrowing, as well as future output and capital values.

Very high rates of growth in the property sector are often followed by severe adjustment. 'Property bubbles', for example, can occur where high rates of demand are fuelled by cheap and readily available credit. In such instances financial institutions can issue too much loan finance which can turn into bad debt and the demise of some banks. A resultant reduction in lending by the financial institutions can reduce the level of investment into property and can therefore fuel the dramatic downturn of many markets. Indeed, government may compound the issue by raising interest rates to deter a repeat of dramatic cyclical activity.

As property markets become depressed the property cycle enters a downward phase. Interest rates are likely to be high leading to business failure and the vacancy of some property. The excess supply of property can be further exacerbated as new development initiated at an earlier, more buoyant time, reaches the market. As a consequence both rental and capital values decline. In the main new development activity would come to a halt although developments already underway continue as they have already consumed substantial costs.

Once confidence returns, the cycle may begin its upward path once again. Initially appropriate and usable excess capacity will be taken up. However, as a period of time has, by definition, taken place since the last recovery much of the vacant building stock may be in too poor a state of repair to be readily used or it may not accord with changes in technology that have taken place. Moreover, before confidence is fully restored entrepreneurs may initially be tempted to overuse their existing space.

Whereas much of the above debate concerning property is concerned with the amplitude of the property cycle it is important to note that the wavelength of the cycle may partially be determined by the following factors:

- the length of the development process
- valuations based upon past and present values and rarely future ones and can therefore be a barrier to change
- likewise developers normally examine past and current values and are not generally speculative
- compounding the above, lenders use values and developers' appraisals in determining the risk of any loan against property
- agreed rent reviews and leases exist and therefore changes in values are not fluid
- many aspects of the property industry have traditionally been accused of conservatism as key players seem slow to respond to new information and ideas.

A low and steady rate of inflation

Inflation is the term used to describe the process whereby the general level of prices rise in an economy. It is normally expressed in the form of an annual percentage change. Thus, an inflation rate of 10 per cent implies that, on average, prices in the economy in question are 10 per cent higher than they were a year ago. Typically, inflation confronting consumers is measured using a retail price index (RPI). Such a price index is a measure of how prices change from month to month for a typical basket of goods that are purchased by the average family. Items in the sample basket are then weighted to statistically reflect the importance and relative use of each item.

The figures produced by such a measure ought to be treated with great caution. As with any average they should only be used as a general summary guideline. The reasoning behind this statement is that the figure will only be totally accurate if one is examining the average householder who purchases exactly the same bundle of goods as those selected by the statisticians calculating the RPI. Obviously, great variations do occur, and more detailed measures of regional rates of inflation, or rates affecting different income groups, or industries, may be more appropriate. International comparisons are also potentially misleading as some countries select different items than others in the production of their inflation statistics, and alternative weightings may be placed upon the importance of certain items. Such variations are bound to occur as consumer needs differ around the world. For example, in cold climates expenditure on heating is likely to be a large and important part of the consumer budget, whereas such an item is mainly irrelevant to the consumer living in a country with a warmer climate. Even looking at inflation over time within a country can be misleading, as the make-up and weighting of the representative bundle of goods can also alter as consumer preferences change. In the property arena it is usually possible to obtain separate information relating to the rate of price changes for specific areas of the built environment. Figures are available, or can readily be calculated, to show building cost increases, house price changes and commercial rental growth, for example.

The goal of low and steady inflation is again largely designed to produce confidence in the economy. For this goal to be achieved the rate of inflation should consistently fall within declared government targets. If this were to be the case, a property investor would be able to assess with a degree of certainty the likely value of a future stream of returns from an investment after the erosive power of inflation has been taken into account. High or volatile rates of inflation will obviously make such calculations more difficult and less certain. As a consequence investment becomes increasingly subject to risk and is therefore less attractive. A fall in the level of investment will reduce the rate of growth in the economy (see the debate above regarding economic growth in addition to the discussion on the multiplier effect in Chapter 10).

Arguably a stable rate of inflation is perhaps more beneficial to an economy than having no inflation at all. The rationale behind this point of view is as follows. If prices consistently rise by say 3 per cent per annum, and if producers cannot pass all of this increase on to the consumer, they will have to devise more efficient methods of production in order to maintain their profitability levels in the light of rising input prices. (See Chapter 6 where the debate concerning a firm's ability to pass on a rise in tax to the consumer is a potential parallel as it is an example of rising costs.)

If a low level of inflation is not achieved and price rises exceed increases in wages people will experience a decline in purchasing power as their real incomes decline. A reduction in spending power will lead to lower levels of demand for goods and services (see Chapter 1 regarding the importance of income in the demand function). Due to the workings of the negative multiplier there will be a detrimental impact upon the fortunes of the property and construction sectors of the economy. In summary, a reduction in income will lead to less spending which in turn will lead to fewer new buildings being required as the rate of economic activity declines.

The direct impact of inflation upon construction firms is that building cost increases pushes up operating costs that are unlikely to be mirrored by increases in revenue. In fact revenues are likely to fall, as the level of demand is more likely to be falling rather than rising. Such a resultant reduction in profitability can mean that marginal projects are not undertaken and firms involved in such activity could go out of business. However, if a firm has already commenced a project and higher rates of inflation take hold during construction, it is probably best to attempt to complete the development and sell it in order to recoup some costs. Otherwise the firm will be left with a partially completed site and no revenue. In the same way, forecasts of escalating inflation may encourage a developer to build more quickly so as to reduce the increased exposure to uncertainty over time.

Finally, inflation can also lead to a redistribution of income. Typically the better off in society are in a stronger position to defend themselves against inflation. Those on higher incomes may hold managerial positions in firms with the capacity to award themselves real salary increases. In addition, the well-off are more likely to hold savings that can, in the short run, be used to cushion the effect of rising prices. As such one may witness the continued development of prestige offices and luxury homes, yet at the lower end of the income scale, less people will be able to afford even a basic starter home.

A low level of unemployment

Unemployment is a measure of the number of people of working age who are officially defined as being without work. Just as with inflation, unemployment statistics need to be analysed with great care. For example the definition of what is meant by the term unemployment can differ between countries and has even been changed over time within countries themselves. Moreover, many who may not wish to register as unemployed will not appear in the official statistics, and as such government data may be an understatement of the actual reality.

If there are a large number of people unemployed, they will be on low incomes and therefore there will subsequently be a lower rate of economic activity than would have been the case if all were in productive employment. This reduction in national income will initiate a negative multiplier process that will have a detrimental impact upon the construction industry and activity in property markets in general. Moreover, unemployment is frequently a regional feature with higher rates experienced in depressed regions as opposed to those seen in more prosperous areas. Therefore, it is not uncommon to witness large volumes of construction activity in one area, yet blight and dereliction in another. Generally though, high levels of unemployment lead to a reduction in incomes and a corresponding decline in construction orders. Because of the labour-intensive nature of the construction process, the building industry typically suffers unemployment rates that are significantly higher than those for the economy on average.

A balance of trade

A balance of trade occurs when a country manages to pay for all of its imports from its earnings from exports. Temporary surpluses or deficits are not seen as a major problem. However, if either condition is allowed to persist difficulties may arise. If a long-run deficit were to occur it implies that monetary outflows from the economy are greater than corresponding monetary injections. In other words imports exceed exports. As consumers spend more on foreign goods, the demand for locally produced output falls and therefore a decline in the domestic manufacturing sector is inevitable. Initially this will lead to a reduction in construction orders in the manufacturing sector, however declining orders could be experienced in all sectors as a negative multiplier effect is initiated. For example, if domestic firms laid off workers the level of unemployment would rise leading to reductions in demand in both the retailing and residential sectors. Eventually such a balance of trade deficit would have to be corrected by government if the country were not to fall further into debt.

On the other hand, a persistent surplus on the balance of trade account can also be detrimental to the economy's performance in the long run. An economy in surplus is likely to stimulate a favourable image of strength in the economy in the eyes of the external investor. Therefore it is likely to attract an inflow of investment funds into the country. This injection of money will tend to alter the exchange rate by increasing the value of the internal currency in relation to other currencies. Effectively, the value of the domestic currency rises as the demand for it goes up due to purchases by foreign investors.

As the value of domestic currency increases the construction industry can be affected in at least two ways. The first impact is that imported construction materials and services now become relatively cheap as more can be purchased from abroad due to the relative strength of the domestic currency. In such situations, therefore one should expect an increasing number of foreign items in domestic buildings, as overseas products become more price competitive with local products. For example, imagine that the local currency is denoted by the symbol '£' and that the currency of the leading trading partner is denoted by the symbol '$'. If originally the exchange was £1=$1 one would need £100,000 to purchase $100,000 of construction materials from abroad. However, if the value of the domestic currency was to rise so that the new exchange rate was £1=$2 one could now purchase $100,000 worth of construction materials for only £50,000, thus making the imported goods much cheaper than before. The problem with this is that domestic building materials suppliers will lose orders and this is likely to induce a negative multiplier effect. The second impact upon the construction industry is that a rising exchange rate will reduce its export potential. For example, at the initial exchange rate of parity, an overseas project yielding a profit of $1 million abroad will produce a profit of £1 million at home. However, as the value of domestic currency rises, the value of the overseas profits will drop when they are repatriated to the home country as $1 million will only be worth £500,000. Therefore, under such circumstances many overseas projects will become unattractive investment proposals.

It can now be seen that a variety of important economic objectives need to be pursued in order to achieve a healthy economy. The welfare of the construction industry, and those industries related to it, whether one looks at chartered surveying or the building materials supply industry as examples, are inextricably dependent upon the fortunes of the economy itself.

12 Macro-economic management, property markets and the construction industry

All economies tend periodically to suffer from periods of recession, when the rate of economic growth is low and unemployment is high (see the discussion on business and property cycles in Chapter 11). However, times of poor economic performance are then normally followed by periods of recovery and relative prosperity. These high points in the business cycle can lead to escalating inflation and rising imports, if the economy begins to 'overheat'. This process tends to repeat itself over the years so that the economy proceeds, usually with some underlying growth, in a cyclical manner. This cyclical path of economic activity, and the terminology used to describe its different stages, is shown in Figure 12.1.

In Figure 12.1 it can be seen that when an economy is experiencing a recession the level of economic activity falls, whereas a depression is where the economy reaches the very depths of a recession. When the economy experiences more rapid growth after a depression it is said to be in a recovery, with the very peak of such a recovery being known as an economic boom. The average rate of growth that the economy achieves is given by the growth path drawn as a line of best fit through the ever-changing cycle. As

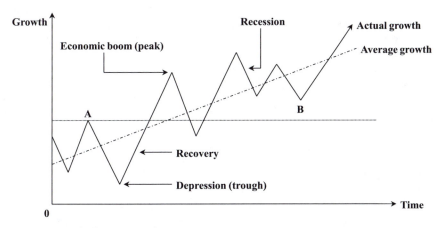

Figure 12.1 *The business cycle*

already implied, it should be noted that either extreme of the cycle, 'trough' or 'peak', may be disadvantageous to the health of the economy. Moreover, an economy that exhibits rapid and unpredictable fluctuations may deter investment in that economy as investors are unsure of where key variables, such as the level of demand, will be in the future.

In order to attempt to achieve a stable and steady rate of growth, the government has a variety of economic management tools at its disposal to attempt to reduce both the frequency and amplitude of such cycles. By managing the economy effectively the government would hope to promote a strong, confident economic environment.

Such intervention in the macro-economy by the government can be via the use of either demand management policies or supply-side policies. Both of these techniques of cyclical management are discussed in detail below. In the following analysis it should become apparent that the construction industry itself can be targeted and manipulated as a tool of macro-economic economic policy. Because of the size of the construction industry, and the nature of its product, inasmuch as most economic activity takes place in buildings, it is easily affected by such policies.

In reality though, it must be recognized that the impact and effectiveness of much economic policy is often severely reduced as some decisions are made for political reasons rather than purely economic ones. In fact it can be argued that many government actions are solely motivated for political gain. Such a political emphasis can eventually cause long-run damage to the economy in exchange for brief, vote catching, short-run economic successes.

Demand management

Demand management is where the government tries to manipulate the level of aggregate demand in the economy. The two main policies that are available for such demand management are fiscal policy and monetary policy. These policies can be used independently, but it is more likely that they will be pursued in conjunction with one another in order to achieve overall economic objectives. The chosen emphasis upon either fiscal policy or monetary policy will depend upon the ideology of the government using the policy, as well as the task that demand management is being used for. The reasoning behind this statement should become apparent in the following analysis as fiscal policy and monetary policy are examined in detail. Reference to both Figures 10.1 and 12.1 will help an understanding of how these policies can be used. These two diagrams demonstrate how movements in the aggregate demand schedule follow through to the economic cycle.

Fiscal policy

Fiscal policy is a demand management technique whereby governments directly intervene in the running of the economy. Fiscal measures involve changing either the level of government expenditure or the level of taxation.

As this is a relatively direct approach to economic management, fiscal policy is seen as being a more interventionist policy than monetary policy. It is for this reason that a reliance on fiscal policy is often associated with governments that are to the left of the centre in political terms. However, it should be noted that most governments, whether on the 'left' or 'right' of the political spectrum will use a degree of fiscal policy in their management of the economy. Fiscal adjustments of government expenditure and taxation so as to fine-tune the economy are typically announced in annual budgets by the minister of finance, or equivalent (chancellor).

In summary, fiscal policy can be seen to have two instruments:

$$\text{Fiscal policy} = \Delta G \text{ and/or } \Delta T$$

Where: Δ = change in
      G = government expenditure
      T = taxation

The following analysis examines how fiscal policy can be used at various stages in the economic cycle, with specific reference to the potential impact of the policy upon the construction industry and the property market.

Economic recession

If the economy is experiencing a recession, or worse a depression, aggregate demand will be low and could be represented by the aggregate demand curve AD_1 in Figure 10.1 and Figure 12.2. This situation of low aggregate demand is normally referred to as a deflationary gap. That is, there is a gap between where aggregate demand should be for a healthy economy, and where it actually is. Such depressed demand could have been induced by an

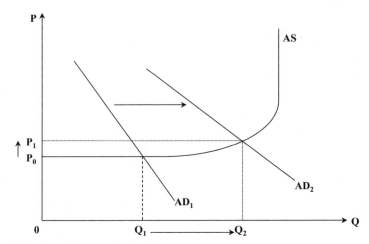

Figure 12.2 *Expansionary fiscal policy*

escalation of world energy costs, for example, leaving consumers and firms with little spending power or investment finance.

As the level of aggregate demand declines, fewer transactions take place in the retailing, office and industrial sectors of the economy. Reduced activity in these sectors is likely to lead to people becoming unemployed and some firms going out of business. Therefore, in terms of the built environment, there is unlikely to be a great deal of new construction. Rather, the slowdown in business is likely to lead to vacant or even derelict commercial buildings, and a fall in real rental values as demand declines. In the residential market, house prices are likely to drop as incomes decrease. Indeed those who have been made redundant, and who did not take out mortgage insurance, may be unable to keep paying off their mortgage and thus may end up having their housing repossessed. In very severe recessions, many may be unable to afford even basic repair and maintenance expenditure on their buildings that could lead to an unsafe and unsightly building stock.

If the economy reaches such a situation, the nation will under-utilize its resources, and suffer from the social cost of unemployment. In addition its urban areas will experience the visual pollution and related social problems of poorly maintained or unused buildings. These problems tend to lead to further difficulties such as an increased crime rate and other more serious urban disorder. Therefore, governments can attempt to improve the situation by initiating an expansionary fiscal policy aimed at increasing the level of economic activity and overall prosperity. These policies are hoped to have an inflationary effect on the level of aggregate demand management. Such an expansionary demand management policy could be achieved by increasing government expenditure or decreasing taxation. Both of these instruments will now be examined in order to assess their effectiveness in promoting the desired economic growth.

Increasing government expenditure

Generally, if a government increases its expenditure it will normally lead to more public sector projects being undertaken. If the public sector has to employ more people, or contract the work out to private firms, in order to achieve such projects, both the people and the firms will experience rising incomes or profits. Therefore, one would expect the positive multiplier process to be initiated. For example, a building firm contracted to build a hospital will take on more labourers, or employ more subcontractors, and order the necessary materials and plant for construction. Thus, unemployed workers may be put in productive jobs, they will spend more in the shops, the shops will need to order more stock from industry, and so on. Likewise, the materials supply sector may have to increase production, especially if the building of this hospital, in our example, is just a small part of overall increases in government expenditure on a larger building programme. Therefore, the materials supply industries will take on more workers, and the various stages of the positive multiplier process will again be set off.

To analyse the implications of increased government expenditure in more depth it is important to distinguish between two categories of public sector spending. Firstly, there is capital expenditure. Capital expenditure is money spent on new construction work such as roads, schools and hospitals as per the example directly above. Capital grants for buildings in the private sector are also found under this classification. Such grants may be issued to private firms to help them with, or completely cover, the cost of setting up a factory, for example, in an economically depressed part of the country. However, it should be noted that not all capital expenditure is money spent directly on the built environment. Capital expenditure can often include government spending on new machinery and vehicles for example. Despite this, though, it is clear that increased capital expenditure by government will lead to more work for construction firms in the public sector.

The second type of government spending is current expenditure. Current expenditure is money spent on the running of government, the civil service and any nationalized industries. Thus, rather than being individual payments for one-off projects, of which new construction is one example, current expenditure is concerned with meeting ongoing costs. However, it is still an important source of potential work for the construction industry, as current expenditure can include for example:

- Money for housing subsidies and improvement grants. These are often made available to people who own homes that do not conform to current, acceptable standards. For example, houses that suffer from damp or structural problems, or lack basic amenities such as an indoor bathroom, may be eligible for such a grant.
- Money for the maintenance of local authority buildings such as hospitals and schools.

In fact, even the ongoing payment of public sector employees, as part of current expenditure, can have a positive, yet indirect effect upon the level of building work. For example if civil servants receive a large pay award they are likely to spend more in the shops, and the positive multiplier will again be initiated.

Decreasing taxation

Just as with increasing government expenditure, the lowering of taxation is an attempt to stimulate higher levels of aggregate demand. Here one must remember that taxation is levied upon both consumers (income tax and sales tax, for example) and firms (corporation tax and business rates, for example). With respect to consumers, if taxes are reduced they will have more money left over (disposable income) to spend on goods and services. Thus, depending upon their marginal propensity to consume, a proportion of this extra disposable income will be spent on additional consumption again giving rise to the positive multiplier effect. Similarly, if firms are not so heavily taxed, their profits will be higher. Such enhanced profitability may encourage them to undertake projects that were previously deemed to be marginal. For example, the high costs of developing a particular site may have put off a firm wishing to build upon it. However, lower taxes could partially offset such high costs, enabling an acceptable profit to be made by going ahead with the development. Therefore, more commercial activity is likely to be undertaken, again stimulating the positive multiplier process.

Therefore effective expansionary fiscal policy, whether achieved by increasing government expenditure or decreasing taxes, should, in both theory and logic, enable the government to stimulate the economy out of recession. Success here would lead to a shift in aggregate demand from AD_1 to a higher level such as AD_2 in Figure 10.1 or 12.2. At higher levels of aggregate demand there would be a fuller utilization of the nation's resources (given current technological and economic conditions). This can be seen diagrammatically as output reaches Q_2 at an acceptable rate of inflation as the general level of prices has only risen to P_1.

As an economy emerges from a recession, general aggregate demand increases causing the aggregate demand curve to shift to the right. This effect, stimulated by the positive impact of the multiplier process, is likely to lead to an increase in building orders in all sectors of the property market ranging from extensions to new build. For example, higher consumer demand could lead to more people working in the retail sector, offices and factories, and therefore there is likely to be a demand for more building space to be created as existing space becomes fully utilized. In the housing market one would also expect an increase in activity as the average consumer becomes better off. However, because there is a time lag involved in the production process, due to the time taken from site acquisition to completion, such increases in demand are initially likely to put pressure on existing space. This will tend to push up rentals and prices as people compete for the available buildings, until new ones become available.

An overheating economy

If an economy begins to overheat it is normally because the growth in aggregate demand has outstripped the economy's ability to supply the increase in the level of economic activity. The economy is liable to overheat when it has reached the peak of a cycle as seen in Figure 12.1. Here, aggregate demand has reached very high levels such as that shown by the aggregate demand curve AD_3 in Figure 10.1. Once an economy has reached this stage, suppliers cannot cope with further increases in orders. Therefore, any additional increases in demand are likely to lead to inflation as prices rise. Alternatively, a balance of trade deficit occurs as imports are used to accommodate the excess demand. Referring to the first point, this situation of excess demand is often referred to as an inflationary gap. Inflation is likely to occur as price rises are used as a way of rationing out existing supply to the highest bidder, whereas an increase in imports could occur as they become comparatively cheap and more accessible. Neither situation of high inflation or an imbalance on the trade account is desirable in the long run. Moreover, if demand were allowed to increase yet further, say to AD_4, these problems would increase. Therefore, government could attempt to dampen down such high levels of demand by introducing deflationary fiscal policy. The same instruments will be used as with expansionary fiscal policy, but in the reverse direction.

Decreasing government expenditure

By decreasing government expenditure there is usually a decline in government capital expenditure rather than a reduction in current expenditure. The reason for this preference of cutting capital expenditure rather than current expenditure is that the latter will tend to lead to more job losses than the former and is therefore deemed to be politically sensitive. Moreover, many may be unaware of proposed government spending plans and are therefore oblivious to the fact that they have been shelved or scaled down. However, as the government cuts back on projects, for example an extensive hospital building programme may be reduced to the provision of new hospitals only in the most needy areas, less people would be employed by the public sector. As a consequence fewer firms will receive public contracts, and therefore general incomes and profitability are likely to fall. If this does occur, those who have lost their jobs will spend less in the shops, and thus the retail sector will order less from suppliers. Therefore, it follows that retail outlets, offices and factories may make people redundant, thereby exacerbating the problem of unemployment via the creation of a negative multiplier. Reductions in current expenditure, say by scaling down the size of the civil service, would also lead to a negative multiplier effect, although the impact on the construction industry would not be as direct as the lost orders caused by reductions in capital spending.

Increasing taxation

In order to promote a similar deflationary effect to that of decreasing government expenditure, the government could attempt to reduce aggregate demand by increasing the levels of taxation. This policy of raising taxes would leave consumers and firms with lower disposable incomes, and as such, they would demand less. Thus, a negative multiplier would again be initiated, as well as a contraction in the magnitude of any positive multiplier effects in the economy.

Therefore, effective deflationary policy, whether achieved by lowering government expenditure, or raising taxation, should be able to revert the economy from inflationary peaks back to a more steady level of growth. However, it must be appreciated that both deflationary and expansionary fiscal policy can fail. Even if this failure is not total, fiscal policy may not reach its desired objectives in full, or it can produce other problems. An analogy can be drawn here with medicines that are prescribed in an effort to cure an illness in a person. First the drugs may not be fully effective on their own and may therefore need to be used in conjunction with some other form of treatment. Second, they may have unpleasant side-effects that also need to be counteracted. To appreciate this point yet further, the potential sources of failure that are inherent in fiscal policy are now briefly examined.

The limitations of active fiscal policy

The potential limitations of fiscal policy are numerous. However, techniques can be devised to limit, or even eliminate, some of these typical problems.

Time lags

Time lags exist because there is a delay between the implementation of a policy and its final results. In other words, the net impact of a policy may occur some considerable time after it, and any associated measures, have been introduced. Such time lags can be divided into two categories, namely decision lags and execution lags.

With respect to decision lags, policy may be delayed simply because it takes time to recognize that the economy is in an undesirable situation. For example, it can take six months or more to collect reliable statistical trends concerning inflation and industrial output. Moreover, before taking action, and deciding upon its magnitude, government must decide whether the upturn, or downturn, of aggregate demand is the beginning of a major change in the economy, or just a minor 'blip' before it settles down to its normal growth path again.

Execution lags, on the other hand, occur because it can take some time for the full effects of fiscal policy, once initiated, to be worked through. For example, long-term spending plans, such as those involved in the construction of new hospitals, cannot be changed overnight. Moreover, it can take a long time to completely implement such a policy. The hospitals need to be planned, suitable sites found, buildings need to be built, staff need to be recruited, and so on. In fact, this process may be completed several years after the original inception of a policy itself. Furthermore, an even larger impact on the economy is likely eventually to occur through the workings of the multiplier process. Thus, even further effects may be felt both on a wider scale and at a far later stage. Execution lags can be further lengthened due to delays created by unexpected shortages, strike action and political wrangling, for example. All of these problems help extend the time before the policy can be fully implemented.

Because of the existence of such time lags, two major problems can occur. The first problem is that the defect in the economy that the policy is aiming to cure can actually be made worse. For example, if government raised public sector expenditure in an attempt to boost aggregate demand and ward off a recession, it may be the case that other variables of aggregate demand have independently and unexpectedly increased also. Thus, the economy could experience increases in public spending as well as increases in private consumption and investment expenditure. Therefore, because the activities of the private sector have not been adequately forecasted, the economy has ended up with far more of a boost than that initially intended by policy-makers. In this situation, the economy could move from the difficulties of a recession to the problems of 'overheating'. In other words, it rapidly swings from one extreme to the other. This scenario could be represented in Figure 10.1. Here a government can attempt to increase aggregate demand from AD_1 to AD_2 or AD_3. However, because of the increased and independent activity of the private sector, aggregate demand could end up as high as AD_4. Such a high level of aggregate demand has an inflationary impact upon the economy with prices rising from P_0 to P_3.

The second problem caused by time lags is that if the government felt that the economy only needed a small boost in aggregate demand, in order to prevent it falling below the desired level of growth, it may only sanction a correspondingly small increase in government expenditure. However, if this rise in public spending was unexpectedly accompanied by a dramatic collapse in investment expenditure from the private sector, the overall impact on the economy could be nullified, or indeed reversed. Therefore, far from aggregate demand moving from AD_1 to AD_2, it could actually drop to a lower level such as AD_0.

Uncertainty

The government faces two major sources of uncertainty when deciding when best to implement fiscal policy, and to what degree. Firstly, it cannot know for certain the value of key variables in the economy such as the marginal propensity to consume. Without accurate data on the marginal propensity to consume, for example, the exact magnitude of the multiplier cannot be known. Such information can only be gathered from past data and informed forecasts. Mistakes in assessing these values could lead to the aggregate demand schedule shifting by too much or too little. For example if the government believed that the magnitude of the multiplier was in order of 2.5, it would expect overall incomes to increase by two and a half times after any injection in government expenditure. Therefore, if, in reality, the multiplier were then found only to be of the order of 1.5, the initial increase in government expenditure would not have been sufficient to reach the desired objectives of the policy.

The second area of uncertainty is that, as suggested above, the government is unsure as to whether other variables, such as private consumption and investment, have moved independently whilst the policy takes effect. If other components of aggregate demand change considerably the net effect of the policy may be quite different to what was originally anticipated.

Fiscal drag

It has been found in the past that much of the positive momentum created by an expansionary fiscal policy can be reduced via the process of fiscal drag. The difficulty occurs as expansionary policy increases the level of incomes in the economy. This is not in itself a problem, but as the incomes of consumers increase, aided by the multiplier process, those in employment can be pushed into higher tax brackets. This eventuality is especially acute under a highly progressive tax regime. Therefore, the full impact of an increase in incomes can be reduced as much of any additional income is then taken back in the form of higher taxation. Thus, there is no great increase in disposable income and aggregate demand.

Opposing multipliers

Government must be careful to ensure that if they wish to create growth in the economy, for example, they are not, at the same time, initiating policy that could inhibit such growth. For instance, by looking at the 'balanced budget theorem' it can be seen that the full effect of an injection of government expenditure, and its associated positive multiplier, can be severely reduced by a corresponding leakage and its associated negative multiplier. Such a leakage would occur, for instance, if taxation were to be raised to pay for the above increase in government expenditure. For example, if the government wished to spend £200 million on a new road-building project, and assuming a marginal propensity to consume of 0.75, the positive multiplier formula would suggest that overall incomes in the economy would be boosted by £800 million. This can be shown in numerical form as:

$$Ku = \frac{\Delta G}{(1 - mpc)}$$

$$Ku = \frac{£200m}{(1 - 0.75)} = + £800m$$

However, if government were to raise taxation by £200 million in order to pay for this increased level of public spending, the impact of the corresponding negative multiplier created by the tax increase would also have to be calculated:

$$Kd = \frac{- mpc \times \Delta T}{(1 - mpc)}$$

$$Kd = \frac{-0.75 \times £200m}{(1 - 0.75)}$$

Therefore, the net effect of the policy will be the difference between the positive effect of the upward multiplier, and the negative effect of the downward multiplier. Thus, in this case, overall incomes will only have increased by £200 million as seen by the calculation below.

$$\frac{\Delta G}{(1 - mpc)} + \frac{- mpc \times \Delta T}{(1 - mpc)} = \frac{£200m}{(0.25)} + \frac{-0.75 \times £200m}{(0.25)}$$

$$= + £800m + (-) £600m$$

$$= + £200m$$

In the same way many of the benefits of increased incomes, brought about by government spending, can leak abroad to other countries especially if there is a high marginal propensity to import.

The failure of pump priming and the theory of crowding out

Many advocates of fiscal policy believe that a small increase in government expenditure will encourage an increase in investment expenditure from the private sector. The theory behind this view is that government spending will promote growth in the economy, via the multiplier process, which should stimulate the confidence of investors to expect a more buoyant market. Therefore, as the public sector spends more, it is hoped that businesses will increase their levels of investment so as to prepare for the anticipated increase in sales volume. This notion of stimulating private investment via an initial injection in government expenditure is known as 'pump priming'.

However, empirical evidence has shown that such pump priming has frequently been associated with decreases in private sector investment expenditure rather than increases. In fact, it is often argued that such a policy 'crowds out' private investment. In its simplest form, crowding out can be explained by the fact that if governments undertake major development projects themselves, with their own organizations, there is less of such work available for the private sector. For example, a private sector house-building firm may shelve plans to build low income housing in a particular location if the government decides to build public sector housing in that area. In this instance, much of the market will have been accounted for by the new public sector housing, thus making it harder to sell the private houses. In other words, the private development is likely to become less viable and potentially unprofitable. Therefore, the necessity for private sector involvement in the market has been effectively crowded out by public sector activity. However, a more technical explanation and form of crowding out can be seen in the following analysis.

If the government borrows the money it requires for increasing its expenditure, this could lead to an upward pressure on interest rates. Higher interest rates are likely to make it too expensive for firms in the private sector to acquire borrowed funds for investment finance. Interest rates could rise for two reasons. Firstly, if the government borrowed money from the financial institutions, they would be in competition with private borrowers for the available funds from such institutions. As such, the financial institutions could ration out their limited supply of loanable funds by using the price mechanism. This would be done by raising the price of finance (interest rates) so that the loans went to the highest bidder resulting in the financial institutions reaping the highest possible reward (profit) from their lending activities. A second possible explanation of rising interest rates would be if the government borrowed directly from the public. Such borrowing could be achieved via the issue and sale of government bonds, for example. In order to persuade the public to put their money in such assets they would need to be offered an attractive return in terms of both security and profitability. If better terms were offered by the public sector, investors may begin to withdraw their money from financial institutions, such as the banks and building societies. In order to maintain business, these institutions would have to react in order to attract funds back to them by giving greater incentives than those being offered by the public sector. Thus, an increase in government spending that is fuelled by borrowing, can lead to higher interest rates.

International research has shown that the problem of crowding out certainly exists but that it is not total. In other words, the fear that an increase in public sector expenditure would be matched by a corresponding and equal decline in private expenditure seems to be unfounded. However, these empirical studies are often classic examples of poor science. The reasoning behind this accusation is that many researchers fall into the trap, or are obliged to do so, of proving their view rather than merely testing it to see if it is correct or not. Consequently there exists an enormous range of statistical findings. Research from people who advocate a prominent role for fiscal policy in the management of the economy, suggest that crowding out may exist but it is of an insignificant magnitude to be problematic. On the other hand, studies by those who believe in a less interventionist approach to demand management show that the magnitude of the crowding out problem is so great that it could totally nullify the usefulness of fiscal policy.

Conclusions on fiscal policy

As shown above, there are a variety of potential problems with fiscal policy. In fact, some argue that the adherence to this policy as the mainstay of demand management in the past has actually made the cyclical nature of the economy worse rather than better.

However, it would be naïve to progress this argument to conclude that fiscal policy should be completely abandoned, as many of the criticisms cited above could be reduced. For example, time lags could be shortened via improved forecasting, having off-the-shelf plans for public buildings, designating public sector land for immediate development when required, and so on. All of these measures would help to facilitate a more rapid implementation of an expansionary fiscal policy.

Moreover, as one cannot rewind and replay history, nobody can possibly know, with any great certainty, what would have happened to world economies if fiscal policy had not been used. For example, if economies had been completely left to themselves the cycles could have been even greater or more frequent. Furthermore, if it was felt that the shortcomings of fiscal policy were too great to use it as a major part of demand management, its role could be suitably reduced to that of merely fine-tuning the economy. Fine-tuning is where small fiscal adjustments are made to keep the economy on course so as to achieve a steady state of growth. Fine-tuning can be a continual process, but such measures are often announced on a country's budget day. On such occasions the chancellor, or finance minister, typically makes relatively minor adjustments to the tax regime and government expenditure in the hope of encouraging the economy to reach government policy targets and maintain a steady course. Finally, if one is not happy with fiscal policy, an alternative area of demand management could be tried in isolation to fiscal policy, or in conjunction with it. The alternative is known as monetary policy.

However, before leaving the discussion on fiscal policy and going on to examine monetary policy, it is felt necessary to provide a brief explanation of the concept of automatic stabilizers. As the name suggests, these are essentially mechanisms that automatically reduce shocks to the economy, and thus the magnitude of cycles, without direct discretionary fiscal policy whereby governments physically intervene in the management of aggregate demand (as seen directly above). For example, if an economy is experiencing a recession, incomes and output will decline. To counteract this decline, injections of money into the system in the form of unemployment benefits, for example, could automatically occur so that even if people were out of work their spending power would not fall to zero. Moreover, in an economy with a progressive tax structure, any fall in incomes could be effectively reduced, as people move into lower tax brackets and therefore pay less tax. In this way, the level of their disposal income is unlikely to fall as dramatically as it would have in the absence of such a tax structure. Both of these effects could help prevent the economy falling even further into recession. Conversely, if the economy was overheating, unemployment benefit payments would fall, and people would move into higher tax brackets. Both of these effects would automatically reduce the potential inflationary effects of such a situation.

Monetary policy

Just as with fiscal policy, monetary policy is a form of demand management in that it is designed to manage the level of aggregate demand in an economy. The main underlying difference is that monetary policy is perceived as being less interventionist than fiscal policy as it is conducted through the intermediary of the financial institutions. It is for this reason that it is often the preferred tool of right-wing governments that advocate less state interference in the workings of the economy. In comparison to fiscal policy, monetary policy is seen as being quicker to react and to implement. However, it must be noted that monetary policy need not be used in complete isolation to fiscal policy as both policies can be used in conjunction with one another in order to achieve desired policy goals.

Monetary policy is a technique that enables government to attempt to manage the level of aggregate demand by either changing the rate of interest or altering the availability of credit.

In summary, monetary policy can be seen to have two instruments:

Monetary policy $= \Delta r$ and/or ΔCR

Where: Δ = change in
$\quad\quad\quad r$ = the interest rate (the cost of credit)
$\quad\quad\quad CR$ = the availability of credit

Please note that although the term 'the' interest rate is frequently referred to it can have different meanings. For example the actual figure quoted may be the official government guide, or base rate, or is an average of interest rates in the economy. This point must be made clear, as there can be a very wide range of interest rates in an economy at any one time for both borrowers and depositors.

Just as with the debate on fiscal policy, the text now goes on to examine how monetary policy can be used to influence the level of aggregate demand at different stages of the economic cycle.

Economic recession

If the economy is experiencing a recession, expansionary monetary policy could be introduced in an attempt to increase the level of aggregate demand, and thus the level of incomes and output. Such a policy would hope to achieve this via a rise in borrowing and spending by both consumers and firms. This could be attempted by lowering the interest rate or increasing the availability of credit.

Reducing interest rates

As many consumers obtain credit in order to finance a proportion of their expenditure, it is likely that a decrease in the cost of such loans would encourage more people to apply for them. As more people obtain loan finance their spending power is increased, and via the multiplier process this is likely to have a positive impact upon the construction industry. For example, more consumption expenditure could lead to the redevelopment of inner city retailing areas, and the setting up of new out-of-town retail parks on the periphery of the urban area. These increases in commercial activity are bound to filter, in part, to domestic industry, perhaps leading to new factory space being ordered. In this process the derived nature of construction demand can again be seen.

There is also likely to be a direct demand for construction in the residential sector as the majority of people finance the purchase of housing through borrowed funds in the form of a mortgage. Thus, if interest rates decline, mortgages become cheaper, and therefore more people can afford to buy their own home. This has the effect of pushing up the demand for housing, and encourages further development. However, the process will eventually come to a halt as house prices are driven up so much by the rising demand that they reach the extent that housing becomes too expensive regardless of cheaper mortgage finance.

In addition to a rise in the demand for housing, if people had to pay less for their mortgage each month, they would have more money left over to afford home improvements and extensions to their existing dwellings. Furthermore, they would have more disposable income for general consumption expenditure. Such increases in general expenditure are likely to filter through to the construction industry via the multiplier process (as described above).

Lower interest rates should also encourage firms to carry out more work and investment. The reasoning behind this statement is that many projects that were previously deemed unprofitable, or too marginal to risk, could now be worthwhile due to the availability of cheaper project finance. Even if the revenues expected from the projects remained unchanged, their cost would now be substantially lower, assuming that a large part of project finance is in the form of borrowed money.

Finally, low interest rates could stimulate an increase in the magnitude of the multiplier process by increasing the marginal propensity to consume. The marginal propensity to consume may increase as the marginal propensity to save declines as the return on savings is reduced via a lower interest rate. This is especially the case if the rate of inflation exceeds the rate of interest. If this situation occurs, the real rate of interest is in fact negative. In other words, the purchasing power of one's money actually declines if it is left on deposit in the form of savings.

Increasing the availability of credit

As with the lowering of the interest rate, the attempt here is to make credit more freely available so as to stimulate aggregate demand. In fact if any government-imposed restrictions are lifted from the financial institutions, credit should become more readily available to those consumers or firms that are in a position to obtain it. Furthermore, in the absence of a dramatic change in the demand for loans, as the supply of loanable funds increases, the price of credit (the interest rate) would fall. Thus, a liberalization of the financial markets is likely to enable more to obtain more loan finance. This should then lead to an increase in the demand for goods and services, thereby creating a multiplier assisted increase in aggregate demand.

As a result, the economy could emerge from the recession, and experience the benefits described earlier after the section on expansionary fiscal policy. However, as with any expansionary policy, there is a risk of the economy growing too fast so that aggregate demand outstrips aggregate supply leading to problems of 'overheating'.

An overheating economy

If an economy does begin to show signs of excess demand, a deflationary monetary policy could be used in order to inhibit demand. Here, government could increase the rate of interest or decrease the availability of credit.

Increasing interest rates

Higher interest rates mean that any proposed consumption, or investment, obtainable from borrowed funds, becomes less attractive, as does the prospect of obtaining a mortgage for house purchase. Moreover, those already with a mortgage, or other outstanding debts, will be faced with higher monthly repayments thus leaving them with less income for the purchase of further expenditure. This is assuming that the loans are normally based upon a variable rate of interest, and that a rate has not been fixed for the duration of the loan. This assumption is reasonable, as although the latter form of loan does exist, the former is still very common.

As a result of increasing interest rates, a negative multiplier effect is likely to be set in motion as fewer goods and services are purchased by consumers and firms. This, in turn, may force shops, offices and factories to cut back output and make staff redundant. Further damage can occur as high interest rates could reduce the magnitude of the multiplier by encouraging a higher marginal propensity to save, and therefore a lower marginal propensity to consume.

Construction firms are adversely affected by such a situation, as they are likely to suffer from declining orders. Moreover, their profitability will decline as they are forced to cut tender prices so as to compete for the little remaining demand. Furthermore, extra costs incurred by unanticipated increases in the interest rate can fatally damage projected cash flows. This situation is often made worse as suppliers press for more rapid repayments as the cost of their borrowing increases also. During such periods, many of the weaker building firms could go out of business, whilst others struggle to survive by taking on marginal projects, or by entering areas such as repair and maintenance, or even seeking contracts abroad.

Decreasing the availability of credit

The government could impose restrictions upon the lending institutions in an effort to reduce their ability to lend money. If credit becomes less easy to obtain, consumption expenditure should decline, therefore reducing aggregate demand. However, it should be noted that such restrictions could also impair the supply side of an economy, as investment expenditure becomes less available to firms as well. The effects of any credit restrictions would be further enforced due to the fact that if credit is successfully limited, interest rates are likely to be forced up as the demand for funds from different sectors competes for the reduced supply of finance. The construction and property industries would both tend to suffer under a regime of credit restrictions, as most activity in these markets is conducted with the assistance of borrowed funds.

Thus, effective deflationary monetary policy should be able to reduce inflationary pressures and encourage the economy to grow at a gentler rate, just as expansionary policy should promote the level of growth. However, as with fiscal policy, there are also some potential limitations of monetary policy. Some of the most important problems are discussed below.

The limitations of monetary policy

Firstly, one needs to examine whether it is in fact logical, or realistic, to expect consumers to react to current interest rates, or whether a longer-term view is actually taken with respect to the cost of finance. Consider the following example in relation to housing. Although a policy of high interest rates is not designed to impair home ownership, housing is an example of a large purchase that is normally financed to a large degree by credit. If current interest rates are higher than usual, this should not necessarily mean that the demand for mortgages, and thus owner-occupied housing, should fall. The logical rationale behind this statement is the recognition that a mortgage is repaid over a long, normally twenty-five year, period, and that interest rates are likely to fluctuate frequently during the life of the mortgage. Therefore, high interest rates merely mean that the homeowner would be temporarily worse off as the level of monthly repayments rose. Conversely, in times of lower interest rates, the homeowner would be better off. Thus, it should be appreciated that periods of high interest and high repayments should be offset by periods of lower interest rates and lower repayments, as interest rates fluctuate around a historic norm.

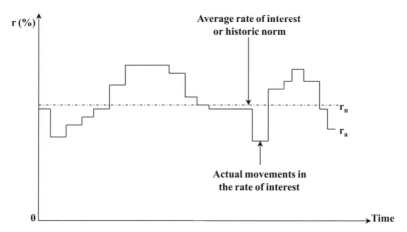

Figure 12.3 *Fluctuations in the rate of interest over time*

Such interest rate fluctuations can be seen in Figure 12.3. Empirical studies have shown that this behaviour does occur in reality and that the demand for loans only begins to fall when the interest rate is so high that the initial monthly repayments cannot be made. In fact, when examining the area of personal loans, for purchases other than housing, evidence reveals that even when charged with very high rates of interest, consumers still take out loans and consequently consumption expenditure is hardly affected. Even those who face higher mortgage repayments in periods of high interest rates can take out a personal loan or second mortgage in order to maintain or increase their spending power. If the interest rate is variable on such loans, part of them may well be paid off when the interest rate is below its historic norm, thus relative savings are made.

A second common criticism of monetary policy is that it is indiscriminate. The argument is that across the board interest rate increases tend to hurt

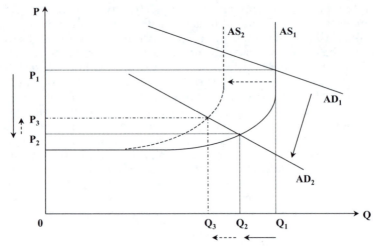

Figure 12.4   *Deflationary monetary policy*

industry as well as consumers. If industry cannot afford to borrow finance for investment, there is a risk that although aggregate demand would fall, and therefore reduce inflationary pressure, aggregate supply would also fall, thus reversing the counter inflationary effect.

This problem can be seen in Figure 12.4. In this diagram we have a starting point of an aggregate supply curve given by AS_1, and a high level of demand in the economy shown by the aggregate demand curve of AD_1. High interest rate policy is then introduced which reduces aggregate demand from AD_1 to AD_2, and thus there is a tendency for inflation to fall from P_1 to P_2. However, high interest rates may cause bankruptcies on the supply side, and make new investment too costly. This would have the effect of reducing aggregate supply from AS_1 to AS_2. Once this contraction in supply has occurred the level of choice is reduced in the economy so that even the lower level of aggregate demand, AD_2, is competing for a smaller number of goods and services, thus driving prices up to P_3. Moreover, such a collapse of domestic firms could be an irreversible feature that leads to the country becoming import dependent. Note that the economy's output has fallen from Q_1 to Q_3.

In practice, a third problem of monetary policy has occurred in relation to the control of the financial institutions. It has been found that when attempting to decrease the availability of credit, as part of a deflationary monetary policy, many governments have had their aims thwarted by the subsequent behaviour of the financial institutions. Financial institutions that have been instructed to decrease the amount of their lending have found ingenious new ways of redefining or packaging their loans so as to circumvent any restrictions. In fact, some have been known to set up separate lending subsidiaries, perhaps abroad, that are outside the control of the initial policy. Moreover, the international nature of financial markets increasingly allows for funds to be raised overseas if they are not forthcoming.

Conclusions on monetary policy

Monetary policy is an alternative method of demand management to fiscal policy, although the two are likely to be used in conjunction with one another as an overall package with which to address problems in the economy. However, it can be seen that because of the indirect nature of monetary policy it may be more difficult to control than the more direct fiscal approach.

A general overview of demand management

The whole principle of demand management primarily rests upon the central assumption that people will react to current economic conditions. Thus, it is assumed that a consumer, for example, will spend according to their present income. Therefore, if government can adjust the level of such income, they should be able to influence the level of current consumer demand. However, this Keynesian view of the economy may not be strictly accurate in modern society. Empirical studies have indicated that people do not necessarily spend according to their current income, but potentially they may make their expenditure plans based upon their perceived future income, or even their previous best income. If either of these two points of view hold in reality, the whole impact of demand management policy based upon its Keynesian foundations will be severely limited.

Milton Friedman put forward his alternative hypothesis known as the permanent income hypothesis in a paper entitled A Theory of the Consumption Function published by Princetown University Press in 1957. Friedman's proposal, which is similar to the life cycle hypothesis of A. Ando and F. Modigliani (*The Life Cycle Hypothesis of Saving: Aggregate Implications and Tests.* American Economic Review, March 1963), suggests that consumption is related more to perceived future income rather than present income. For example, Friedman feels that people will over-consume now if they expect that their future income will be great enough to offset any debts incurred at present. Therefore, at the beginning of one's working life, one could acquire, via loan finance, a house, a car and household goods as the individual could reasonably expect future salary increments and job promotions to pay the debts off in the future. Moreover, ongoing inflation will gradually erode away a proportion of the debt owed.

Conversely, the relative income hypothesis proposed by James Dusenberry (*Income, Saving and the Theory of Consumer Behaviour*, Harvard University Press, 1957) suggests that people's spending behaviour is strongly influenced by their best previous income. Thus, if people's incomes decline, in times of deflationary demand management, or a recession, for example, they will not decrease their consumption by much as they have become accustomed to the standard of living provided by a higher level of income. Therefore, loan finance would be sought, or savings used, in order to defend their previous best consumption position. This could be done in the hope that when incomes rise again, perhaps at the end of the recession or during a period of expansionary demand management, sufficient monies will be forthcoming to pay off the debt. The theory then goes on to suggest that once their original income and consumption pattern has been reached, any further increases in income will lead to a positive jump in consumption expenditure. This process was termed the ratchet effect by Dusenberry. Again, the impact of policies designed to influence current consumption may be severely limited by such behaviour.

Due to these apparent inadequacies of demand management some argue that a greater emphasis should be placed on the supply side of the economy.

Supply-side economics

As the name suggests, supply-side economics is the study of macro-economic management that is aimed at managing the level of aggregate supply in the economy. Essentially, supply-side measures are those that are introduced in order to encourage an expansion in domestic supply. The underlying logic here is that if policy-makers can successfully increase overall domestic aggregate supply, the impact of both a demand-led recession, as well as inflationary periods, in the economy would be reduced.

For example, if the economy is growing too fast and overheating, as shown by the high level of aggregate demand AD_1, and the initial aggregate supply schedule AS_1 in Figure 12.5, prices would be high as shown by P_1. However, if domestic aggregate supply could be increased in such an instance, much of the inflationary effect of excess demand could be reduced. This is demonstrated whereby increased supply levels shown by the aggregate supply function AS_2 means that there will be less competition

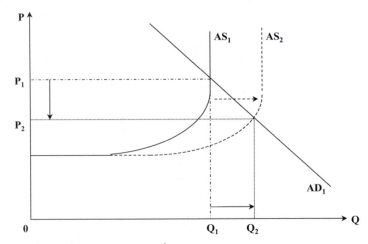

Figure 12.5 *Increasing aggregate supply*

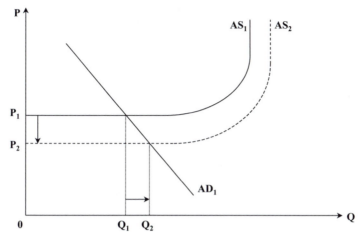

Figure 12.6 *Reducing supply costs*

over the available output so that prices fall from P_1 to P_2, as output increases from Q_1 to Q_2.

Similarly, the promotion of circumstances that enabled aggregate supply to be increased throughout the economy should reduce the decline in output and the level of employment, when aggregate demand is low during a recession. For example, if favourable conditions enabled firms to produce more at any given price, the aggregate supply curve could shift to the right as shown in Figure 12.6. Here it can be seen that at the low level of aggregate demand such as AD_1, output would drop as low as Q_1 under ordinary supply conditions shown by the original aggregate supply curve AS_1. Yet successful supply-side policies could shift the aggregate supply curve to the right thus giving a new curve at AS_2, and therefore output may only drop to Q_2 when demand is as low as AD_1. Under the old conditions of supply given by AS_1, many firms may have gone out of business thus leading to a substantial decline in domestic output to Q_1. However, given more favourable conditions of supply, via the introduction of positive supply-side policies, more firms would have been able to cope with the recession and as such output would not fall as much as would have been the case previously.

Now that the basic principle of supply-side economics has been introduced, the text now examines some specific ideas on how government can aim to boost the general level of aggregate supply in the economy. These ideas can be classified under the heading of supply-side policies, or more meaningfully they should be termed supply-side instruments, as they are each an instrument of the overall policy of supply management.

Income tax cuts

It has been logically hypothesized that very high rates of income tax may act as a disincentive to people who work hard as they will see a large proportion of any additional earnings being taken away in the form of income tax. Thus, lowering the overall level of income tax means higher financial rewards from employment in the form of greater disposable income. This should encourage people to get a job, if they do not already have one, or work harder, and earn more, in the job that they already have. More financial reward from work, in the form of higher take-home pay, increases the opportunity cost of leisure (as this time could be used to earn yet more money), and should encourage people to work harder. Thus, income tax cuts should increase the overall level of labour productivity.

However, in reality, this process may not occur exactly in the form described above. In fact, one may see the reverse effect occurring. Higher financial rewards at work mean that people will have more disposable income and could spend this on improving or increasing their leisure time. For example, the temptation to join the local golf club, and spend more time there may be enhanced if such an activity becomes more affordable. Therefore this potential negative effect on labour productivity could go some way to counteracting the original, positive effect.

Moreover, many people cannot actually work more even if they want to, as they are in 'nine to five' jobs and are paid for that period only. In these cases, no reward would be forthcoming to the individual who worked longer hours. However, it is probably true to say that because disposable income increases with such tax cuts, the general morale of the workforce would be improved making people put more effort into their jobs during the official working day. However, this point leads on to a further problem associated with tax cuts. If people in existing jobs work harder, as suggested above, more work will be completed in less time than was previously the case. Such increases in productivity could reduce the number of vacancies, or indeed lead to redundancies as the work of some is now done by others, thus leading to increased unemployment. For example, a team of bricklayers may normally work on a building site from eight in the morning to six at night. However, if taxes were reduced they may try to work faster and for longer hours so that they could complete their existing job and move on to another site. If this were to happen, the level of competition between bricklaying teams would increase and some would become redundant as others did the work. In fact the benefits of the tax cuts could be quickly eliminated as bricklayers reduced their tender prices in the hope of attracting work in this highly competitive market.

Decreasing the power of trade unions

Some economic commentators perceive trade unions as being out-of-date organizations that have failed to adapt to socio-economic change over time by enforcing outmoded restrictive practices within industries. It is claimed that the effect of this is low output, thus leading to a lower aggregate supply than would be the case if industry were allowed to modernize and operate in a completely unrestricted manner. Moreover, many feel that closed shop arrangements, for example, have kept wages artificially high leading to low labour employment. Trade unions have also been accused of fostering work practices that make domestic firms less competitive than would otherwise be the case. With increased international competition this could lead to the demise of many businesses with an obvious impact upon domestic output and employment.

If the power of the trade unions were to be decreased, or even eliminated altogether, the supporters of the argument above believe that more modern and efficient work practices would be adopted by firms, and wage rates would become more competitive. Thus, it follows that overall domestic supply should be enhanced. However, attempting to introduce such a policy is likely to produce a fierce political debate between those who believe in the benefits of intervention and regulation, and those who believe in promoting a liberalized labour market. For example, property economists who advocate the need for trade unions argue that without trade union support, construction workers could lack job security, and be forced to work in a dangerous environment where an emphasis was placed upon productivity rather than the safety of employees.

Decreasing structural and frictional unemployment

Many countries lose a great deal of potential labour output due to both structural and frictional unemployment. Structural unemployment is the situation whereby key traditional industries have closed down, or have shed significant amounts of labour, putting a large number of people out of work. This is often a problem as many of these workers are only trained to work in their now defunct, or declining, industry. In order to reduce the levels of structural unemployment, it has been suggested that governments should provide retraining programmes so that structurally unemployed labour can work, and produce output, in another industry. Similarly, although strictly not under this category, the provision of training for school leavers could decrease the likelihood of youth unemployment by providing employable skills. To minimize the direct cost burden of these policies upon government, the private sector could be offered tax incentives so as to encourage them to take on young trainee labour, or set up training programmes.

Frictional unemployment is temporary unemployment caused by people who are moving between jobs. Providing labour with assistance concerning such a move can reduce such frictional unemployment. For example, bureaucratic procedures could be simplified and speeded up if they are a contributory cause of any delay. Alternatively, if the delay between jobs is caused by the fact that people need to live elsewhere to take up a new job, a relocation package, offering a range of financial incentives, could help assist their move. Relocation packages can enhance the mobility of labour by helping with costs of moving house like legal fees, estate agents' fees and government taxes on house purchase such as 'stamp duty' (if applicable).

By effectively reducing the number of people who are out of work, the nation's overall output, or aggregate supply, should be enhanced.

Reducing unemployment benefits

Many countries operate social security schemes whereby those in employment pay towards a state insurance programme so that if they, or others become unemployed, there will be a pool of money available from which to claim payments until employment is found again. Some argue that such a system potentially retards aggregate supply on two grounds:

- If one can get money for not working, some will not bother to find work.
- The insurance payments, deducted from earnings, can be perceived as a tax, thus giving people a disincentive to work. This is especially the case where contributions are substantial.

Therefore, the argument here is to either reduce unemployment payments so that unless people want to live at a purely subsistence level, they will need to work, or abolish such a system altogether so that people will have to find work. Unfortunately, however, the obvious faults with both scenarios are:

- There may not be any jobs anyway.
- There may be jobs but people are not trained for them.
- There may be jobs in the prosperous areas of a country, but it may be too expensive to move there, or there is insufficient, affordable accommodation in such areas.

Profit-sharing schemes

Many employees have claimed that they feel alienated at work and therefore owe little allegiance to the firm for which they work. As such, the effort that they put into their work is far from being their full potential. In an attempt to reduce this sense of ill-feeling and low motivation it has been suggested that governments should encourage firms to introduce profit-sharing schemes. The schemes would operate in a way that if output and profits do increase the employees of the firm would benefit by receiving a profits-related bonus to their wages, perhaps in the form of shares. In the same manner, wages could be cut if poor levels of productivity and output lead to losses being made. Such a scheme could be operated by all forms of firms ranging from surveying practices to building companies. Although they have been adopted by some organizations this has been done with varying success. A common criticism of many profit-sharing schemes is that the actual percentage of profits that is distributed to the workforce is so small that it is deemed to be insignificant. Therefore the total impact of profit-sharing schemes upon aggregate supply is likely to be small.

Grants and subsidies

In some instances firms may need an initial incentive to encourage them to set up in business and therefore add to national output. Grants and subsidies could be given to new firms in order to help cover the initial cost of labour or capital employment. Indeed, for an existing firm, a grant or subsidy would reduce the current costs of operation so that it may become profitable to expand output so that again one would see enhanced aggregate supply. However, in reality, the positive impact of such policies only lasts as long as the policy itself. After the termination of benefits, when financial assistance is removed, evidence has shown that one reverts back to the original situation with firms going out of business, or cutting back output, as they cannot survive without public help.

A general overview of supply-side policies

As can be seen from the above, there are arguments both for and against supply-side policies. Which side of the argument one takes in each case depends upon political beliefs just as much as economics inasmuch as many of these points are contentious and emotive issues. Moreover, policies will only be successful if they are introduced intelligently on the foundation of well-researched theory. For example, one must find out whether income tax cuts will lead to people working harder or will give them an incentive to take more leisure time. Both these effects are likely to occur and therefore their relative magnitudes need to be assessed so that one can realistically gauge the likely overall impact of the policy. Furthermore, it is unlikely that supply-side policies can be used effectively in complete isolation from demand management policies. Therefore supply-side economics should be viewed as complementary policy to demand management rather than a substitute for it. With respect to empirical evidence, many supply-side policies have been in operation for such a short period of time that the data available is not yet sufficiently significant to prove either their success or failure.

Local government macro-economic policy and the built environment

It should be noted that in the same way that central government aims to control the national economy, local governments, such as local authorities and town councils, can also attempt to control their own local economies. At the local level a variety of techniques can be used to manipulate the level of investment and consumer spending in order for an area to be directed towards a desired economic objective. For example, if a local authority was confronted with the problem of a declining economy, it could aim to stimulate the economy and initiate a corresponding positive multiplier process in a number of ways as suggested below.

- The giving of low interest loans or capital grants for businesses to set up in the area in the hope of stimulating investment and employment.
- The granting of subsidies to householders in order to enable them to improve the quality of their dwellings to an acceptable standard.

- The spending of public monies on a large project such as an urban renewal scheme with the aim of improving local infrastructure and increasing the attractiveness of the location for potential investors looking for a suitable business location.
- Encouraging firms to locate in the area by offering them a temporary tax holiday from local taxation such as business rates. This would have the effect of lowering the operating costs of the firm in the short run to help counteract high start-up costs, and costs incurred in establishing a market.

The above list of policies is by no means an exhaustive one as it is simply designed to provide an insight into how macro-economic policies can be applied to target the needs of smaller, local economies.

Conclusion

It can now be seen that the fortunes of the property and construction industries are inextricably linked to the performance of the economy in general. Macro-economic theory enables one to forecast the likely impact upon these industries of government policies aimed at tackling macro-economic problems at both the national and local levels. For example, the reader should now have a clear idea of the likely effect on the housing or office market if there were to be an expansionary monetary policy.

Index